MODERN ALGEBRA:
A NATURAL APPROACH, WITH APPLICATIONS

ELLIS HORWOOD SERIES IN MATHEMATICS AND ITS APPLICATIONS

Series Editor: Professor G. M. BELL, Chelsea College, University of London

The works in this series will survey recent research, and introduce new areas and up-to-date mathematical methods. Undergraduate texts on established topics will stimulate student interest by including present-day applications, and the series can also include selected volumes of lecture notes on important topics which need quick and early publication.

In all three ways it is hoped to render a valuable service to those who learn, teach, develop and use mathematics.

MATHEMATICAL THEORY OF WAVE MOTION
G. R. BALDOCK and T. BRIDGEMAN, University of Liverpool.
MATHEMATICAL MODELS IN SOCIAL MANAGEMENT AND LIFE SCIENCES
D. N. BURGHES and A. D. WOOD, Cranfield Institute of Technology.
MODERN INTRODUCTION TO CLASSICAL MECHANICS AND CONTROL
D. N. BURGHES, Cranfield Institute of Technology and A. DOWNS, Sheffield University.
CONTROL AND OPTIMAL CONTROL
D. N. BURGHES, Cranfield Institute of Technology and A. GRAHAM, The Open University, Milton Keynes.
TEXTBOOK OF DYNAMICS
F. CHORLTON, University of Aston, Birmingham.
VECTOR AND TENSOR METHODS
F. CHORLTON, University of Aston, Birmingham.
TECHNIQUES IN OPERATIONAL RESEARCH
VOLUME 1: QUEUEING SYSTEMS
VOLUME 2: MODELS, SEARCH, RANDOMIZATION
B. CONNOLLY, Chelsea College, University of London
MATHEMATICS FOR THE BIOSCIENCES
G. EASON, C. W. COLES, G. GETTINBY, University of Strathclyde.
HANDBOOK OF HYPERGEOMETRIC INTEGRALS: Theory, Applications, Tables, Computer Programs
H. EXTON, The Polytechnic, Preston.
MULTIPLE HYPERGEOMETRIC FUNCTIONS
H. EXTON, The Polytechnic, Preston
COMPUTATIONAL GEOMETRY FOR DESIGN AND MANUFACTURE
I. D. FAUX and M. J. PRATT, Cranfield Institute of Technology.
APPLIED LINEAR ALGEBRA
R. J. GOULT, Cranfield Institute of Technology.
MATRIX THEORY AND APPLICATIONS FOR ENGINEERS AND MATHEMATICIANS
A. GRAHAM, The Open University, Milton Keynes.
APPLIED FUNCTIONAL ANALYSIS
D. H. GRIFFEL, University of Bristol.
GENERALISED FUNCTIONS: Theory, Applications
R. F. HOSKINS, Cranfield Institute of Technology.
MECHANICS OF CONTINUOUS MEDIA
S. C. HUNTER, University of Sheffield.
GAME THEORY: Mathematical Models of Conflict
A. J. JONES, Royal Holloway College, University of London.
USING COMPUTERS
B. L. MEEK and S. FAIRTHORNE, Queen Elizabeth College, University of London.
SPECTRAL THEORY OF ORDINARY DIFFERENTIAL OPERATORS
E. MULLER-PFEIFFER, Technical High School, Ergurt.
SIMULATION CONCEPTS IN MATHEMATICAL MODELLING
F. OLIVEIRA-PINTO, Chelsea College, University of London.
ENVIRONMENTAL AERODYNAMICS
R. S. SCORER, Imperial College of Science and Technology, University of London.
APPLIED STATISTICAL TECHNIQUES
K. D. C. STOODLEY, T. LEWIS and C. L. S. STAINTON, University of Bradford.
LIQUIDS AND THEIR PROPERTIES: A Molecular and Macroscopic Treatise with Applications
H. N. V. TEMPERLEY, University College of Swansea, University of Wales and D. H. TREVENA, University of Wales, Aberystwyth.
GRAPH THEORY AND APPLICATIONS
H. N. V. TEMPERLEY, University College of Swansea.

MODERN ALGEBRA: A NATURAL APPROACH, WITH APPLICATIONS

C.F. GARDINER, B.Sc., M.Sc., F.I.M.A.
Department of Mathematics
University of Exeter

ELLIS HORWOOD LIMITED
Publishers Chichester

Halsted Press: a division of
JOHN WILEY & SONS
New York Chichester Brisbane Toronto

First published in 1981 by

ELLIS HORWOOD LIMITED
Market Cross House, Cooper Street, Chichester, West Sussex, PO19 1EB, England

The publisher's colophon is reproduced from James Gillison's drawing of the ancient Market Cross, Chichester.

Distributors:

Australia, New Zealand, South-east Asia:
Jacaranda-Wiley Ltd., Jacaranda Press,
JOHN WILEY & SONS INC.,
G.P.O. Box 859, Brisbane, Queensland 40001, Australia

Canada:
JOHN WILEY & SONS CANADA LIMITED
22 Worcester Road, Rexdale, Ontario, Canada.

Europe, Africa:
JOHN WILEY & SONS LIMITED
Baffins Lane, Chichester, West Sussex, England.

North and South America and the rest of the world:
Halsted Press: a division of
JOHN WILEY & SONS
605 Third Avenue, New York, N.Y. 10016, U.S.A.

British Library Cataloguing in Publication Data
Gardiner, C. F.
Modern algebra. – (Ellis Horwood series in mathematics and its applications)
1. Algebra
I. Title
512 QA155 80–42229

ISBN 0–85312–285–7 (Ellis Horwood Limited, Publishers, Library Edition)
ISBN 0–85312–303–9 (Ellis Horwood Limited, Publishers, Student Edition)
ISBN 0–470–27115–9 (Halsted Press)

Typeset by Kent Photoprint Limited, Chatham, Kent.
Printed in Great Britain by Butler & Tanner Ltd., Frome, Somerset.

Contents

Author's Preface

For students arriving at university fresh from school two aspects of pure mathematics give particular difficulty, namely the rigour of analysis and the abstract nature of algebra. Recently it was usual to introduce students to abstract algebra very early, often with little motivation. However this fashion is changing. Nevertheless modern abstract methods have great elegance and provide powerful tools for the solution of specific problems not only in other branches of mathematics but also in other sciences such as physics. It is therefore important that students of mathematics should meet these abstract ideas as soon as possible. The problem is to introduce them in a natural way as arising out of situations and problems in which the student is likely to be interested. However it is worthwhile bearing in mind that abstraction is a relative notion. Ideas with which the student has lived for some time seem 'concrete' no matter how abstract they may appear to someone meeting them for the first time.

In this book I have tried to build on ideas of logic, number, and geometry, which the student might be expected to have met at school or in everyday life. Usually ideas like group, ring, and field are introduced as convenient ways of summarising facts that appear in many different forms. Economy of treatment is therefore the initial justification for introducing the abstract ideas at this level. Of course, where feasible, the abstract notions are used to solve concrete problems.

One such unifying concept is that of the vector space together with the associated concept of the linear transformation. We shall use these concepts to give insight into matrices and the sets of solutions of simultaneous linear equations.

Determinants are introduced in a rather old fashioned way using permutations. In my experience the average student seems to find this direct computational approach easier to understand than the various abstract or indirect developments, elegant though these are. Also the computational approach does provide the opportunity of 'slipping in' some work on permutation groups.

The study of quadratic forms is justified as a way of classifying quadrics, which are the natural generalisation of the conics met at school. Simultaneous reduction of two quadratic forms is introduced as a method of attacking vibrational problems in dynamics by finding normal coordinates.

Factorisation of integers and polynomials and the use of this in solving problems in group theory (which provides an excuse for doing a little more group theory) is used as the starting point for an exploration of Euclidean domains. This seems the easiest part of ring theory to motivate. In this way the student can become familiar with some of the basic notions of ring theory, such as prime, principal and maximal ideals, in a relatively painless way.

To make the book more useful to those working largely on their own, I have included outline solutions to all the exercises. The results of the exercises are not used elsewhere in the book.

This book is based on lectures given over several years to first year students specialising in mathematics at Exeter University. Although based on a particular course, it covers the usual topics in algebra which must be studied at some time by both mathematicians and users of mathematics. I hope it will prove useful to all such students, not only for the usual techniques discussed at this level, but also in introducing them to abstract ideas and methods.

Note to the Reader.

Although exercises are given only at the end of each chapter, the reader is advised to attempt them as soon as it is felt that they will illuminate the text and not to wait until the chapter is completed. This advice is particularly important for the very long chapter 4 which I have written as a single chapter in order to emphasise the unity of the material it contains.

Acknowledgement

It is a pleasure to thank Alan Wilkin for his considerable and generous help in providing the excellent figures throughout the text.

CHAPTER 1

The Presentation of Mathematics

1.1 INTRODUCTION

For the moment let us take a naive view of the concept of 'set' and think of a set as some collection of objects. We write $x \in S$ to mean the object x is a member of the set S, and $y \notin S$ to mean the object y is **not** a member of the set S.

In everyday life such sets will usually have only a finite number of members, e.g. a football team. However, in mathematics we are frequently involved with sets which have an infinite number of members, e.g. the set of integers

$$\mathbf{Z} = \{\ldots\ldots -1, 0, 1, 2, 3, \ldots\ldots\}$$

It is with these large sets that difficulties may arise. Around 1903 Bertrand Russell discovered a set which is very large indeed; in fact so large that a contradiction or antinomy arises. This set S, often called the **Russell set** after its discoverer, is defined as $S = \{A : A \notin A\}$. In words this reads : S is the set of all those sets A such that A is **not** a member of A.

Almost every set that we can think of is a member of the Russell set S. For example

$$\mathbf{Z} = \{\ldots\ldots -1, 0, 1, 2, 3, \ldots\ldots\}$$

is not itself an integer, hence $\mathbf{Z} \notin \mathbf{Z}$. Thus $\mathbf{Z}$ is a member of S. The set $T = \{-\frac{1}{2}, 1, \frac{1}{2}\}$ is **not** equal to $-\frac{1}{2}$, 1, or $\frac{1}{2}$, so $T \notin T$. Hence $T \in S$.

Now S is a set according to our naive ideas, therefore it makes sense to ask whether or not S is a member of itself. But if $S \in S$, then $S \notin S$, and if $S \notin S$, then $S \in S$. Thus whichever choice we make, we arrive at a contradiction.

To avoid contradictions like this and to increase clarity and rigour (besides making available economical and powerful abstract methods, as we shall see later), it is the modern custom, pioneered by the German mathematician David Hilbert, to present mathematics in the following way.

(1) Decide on a set of starting assumptions or axioms and make sure, if you can, that these are consistent, i.e. they do not contradict themselves or lead to contradictions. The axioms describe properties of certain undefined objects. In fact the latter are just those 'objects' that satisfy the axioms.

(2) Deduce, by the rules of some logic, further results which are to be considered 'true' when the axioms are held to be 'true'. Of course, in a full statement of the theory, a definition of truth must be given. These further results are usually called THEOREMS and describe properties of the undefined objects described by the axioms.

This approach is called the **axiomatic method**. It is not entirely new. The essence of this approach goes back to the Greek mathematicians, Thales and Euclid, some 2,500 years ago. However, the idea of axiom has changed since then. By axiom *the Greeks meant* "some obvious property of the real world which serves as a starting point for a chain of deduction". By axiom *we mean* "some property of certain undefined objects". In fact, anything will do as our object, as long as it possesses the property specified in the axiom.

Let us consider some axioms in the modern sense and the deduction of a theorem from them. This deduction is called a proof of the theorem. Strictly speaking we should call it a **semi-formal proof**, because the logic is used informally. When we have available enough classical logic for our purpose, we shall return to this point and give a definition of a **formal proof**; see Section 1.3.

Take as undefined objects 'lines' and 'points'. Note that, although the terms are suggestive of objects met in everyday life, as far as we are concerned they are defined only by the properties given them in the axioms.

Take an undefined relation connecting lines and points which we call 'incidence'. When a line and a point are so related, we say they are incident with each other. More suggestively, we say that the point p lies on the line l or that the line l passes through the point p.

All we know about these concepts is described by the axioms as follows.

Axiom A1. For every point p and for every point q, not the same as p, there exists a unique line l that passes through p and q.

Axiom A2. For every line l there exist at least two distinct points incident with l.

Axiom A3. There exist three distinct points such that no line is incident with all three of them.

Definition. Two distinct lines are said to be **disjoint** if there is no point which is incident with both of them.

We now prove the following theorem.

Theorem. If l and m are distinct lines that are not disjoint, then there is one and only one point which is incident with both l and m.

Proof. (1) Because l and m are not disjoint there exists a point incident with both of them; by the definition of disjoint lines.

(2) Assume there are two points A and B incident with l and the same two points are incident with m.

(3) A and B lie on at least two distinct lines, by (2) and the hypothesis of the theorem.

(4) A and B lie on one and only one line, by Axiom A1.

(5) Statements (3) and (4) contradict each other.

(6) Thus assumption (2) must be false.

(7) There are not two points incident with both l and m by (6).

(8) There is at least one point incident with both l and m by (1).

(9) There is one and only one point incident with l and m, by (7) and (8).

Now let us take an example in which we assume that we know the natural numbers and how to factorise them.

Theorem. There are an infinite number of primes.

Proof. Suppose not. Then $p_1, p_2, \ldots\ldots, p_r$ are all the primes. Consider $N = (p_1 p_2 p_3 \ldots\ldots p_r) + 1$.

Now N can be factorised into a product of primes.
However none of these primes can be those in the set

$$\{p_1, p_2, \ldots\ldots, p_r\} \text{ since } p_i \nmid N \text{ for } p_i \in \{p_1, \ldots\ldots, p_r\}.$$

Thus there is at least one prime not in the set $\{p_1, \ldots\ldots, p_r\}$. Contradiction. Thus the initial supposition must be wrong. We conclude that there must be an infinite number of primes.

Note: that this proof is even more informal than the previous one. Not only are we using the logic informally, we are also arguing from a concealed or implicit set of axioms, rather than from an explicit set of axioms describing the natural numbers precisely. We now focus our attention on the logic which we have rather taken for granted up to now.

1.2 CLASSICAL LOGIC

We use the symbols p, q, r,. . . .to denote propositions or statements, i.e. expressions like 'The sun shines', 'All triangles are isosceles', and so on.

The words such as *and*, *or*, *implies*, *not*, etc., are called **logical connectives.** Implication turns out to be not quite what we would expect from everyday usage. In fact, to define our connectives precisely it is convenient to use truth tables. In these tables 1 denotes 'true' and 0 denotes 'false'.

The tables are as follows.

$\wedge$	p	q	$p \wedge q$
	1	1	1
	1	0	0
	0	1	0
	0	0	0

Table 1.2.1 'and'

$\vee$	p	q	$p \vee q$
	1	1	1
	1	0	1
	0	1	1
	0	0	0

Table 1.2.2 'or/and'

$\rightarrow$	p	q	$p \rightarrow q$
	1	1	1
	1	0	0
	0	1	1
	0	0	1

Table 1.2.3 'implies'

$\leftrightarrow$	p	q	$p \leftrightarrow q$
	1	1	1
	1	0	0
	0	1	0
	0	0	1

Table 1.2.4 'if and only if', 'equivalence'

$\sim$	p	$\sim p$
	1	0
	0	1

Table 1.2.5 'not'

As shown in the preceding tables the symbols used for the connectives are: $\wedge$, $\vee$, $\rightarrow$, $\leftrightarrow$, $\sim$.

Any expression built up from propositions using the above connectives in a meaningful way is called a formula.
For example:

$$(p \vee q) \rightarrow ((\sim p \wedge q \wedge r) \vee (\sim p \rightarrow r)).$$

DEFINITION 1.2.1 A **tautology** is a formula which always has the truth value 1 (true) whatever truth values 0 or 1 are given to its constituent propositions or variables.

EXAMPLE 1.2.1 $(p \to q) \to (\sim q \to \sim p)$.

We have the following truth table.

p	q	$p \to q$	$\sim q$	$\sim p$	$\sim q \to \sim p$	$(p \to q) \to (\sim q \to \sim p)$
1	1	1	0	0	1	1
1	0	0	1	0	0	1
0	1	1	0	1	1	1
0	0	1	1	1	1	1

Table 1.2.6

Thus $(p \to q) \to (\sim q \to \sim p)$ is a tautology.

In mathematics we often use expressions involving the phrases 'for all' and 'there exists'. The former is denoted by $\forall$ and the latter by $\exists$.
For example: $\forall x\ (x = x)$, $\exists x\ ((x^2 - 1) = 0)$.
These read 'for all x, x is equal to x' and 'there exists a value of x for which $x^2 - 1 = 0$'.

Of course, for these expressions to have a precise meaning, a domain of interpretation must be specified. That is a domain from which we choose the values of x. For example, in the above we could take our domain to be the set of integers, when both our statements would be true. On the other hand, if we chose as our domain the set of natural numbers $\{2, 3, 4, \ldots\ldots\}$, then the second statement $\exists x\ ((x^2 - 1) = 0)$ would be false. Thus, in general, the truth or falsehood of a statement depends on the domain of interpretation. Statements that are true for all choices of the domain are called **logically valid** and corres- pond to the tautologies discussed earlier before the introduction of quantifiers.

In order to analyse propositions involving quantifiers, let us denote expressions like $x = x$ and $x^2 - 1 = 0$ by symbols such as $P(x)$, $Q(x)$ etc. We can think of $P(x)$ as some proposition involving the variable x. For a proposition involving two variables x and y we use the symbols $P(x,y)$ and so on. We write $P(a)$ for the expression that results when we replace the variable x in $P(x)$ by some specific value a.

For example, if $P(x)$ denotes $x^2 - 1 = 0$, then $P(a)$ denotes the proposition $a^2 - 1 = 0$. In particular $P(2)$ denotes the proposition $2^2 - 1 = 0$; which is false. However $\exists x\, P(x)$ is true, because $P(1)$ is true, so there does exist a value of x for which $P(x)$ holds. (Here we assume our domain contains 1 and 2).

We list now certain rules which tell us how to use quantifiers.

Q1. $P(a) \rightarrow \exists x\, P(x)$.

In other words, if $P(a)$ is true, then there exists a value of x (for example a) for which $P(x)$ holds.

Q2. $(\forall x\ P(x)) \rightarrow P(a)$, where a is any mathematical object in the domain of interpretation.

For example, if we have proved that $\forall x\ (x = x)$ is true, then $a = a$ is true.

Q3. $(\forall x\, P(x)) \leftrightarrow \sim(\exists x\, (\sim P(x)))$.

In words: $P(x)$ is true for all x if and only if there does **not** exist a value of x for which $P(x)$ is false.

Q4. $\sim(\exists x\, P(x)) \leftrightarrow \forall x\, (\sim P(x))$.

This means: there exists no value of x for which $P(x)$ is true if and only if for all values of x $P(x)$ is false.

Q5. $(\forall x\, (P(x) \wedge Q(x))) \leftrightarrow ((\forall x\, P(x)) \wedge (\forall x\, Q(x)))$.

Note that this is not true if we replace $\wedge$ (and) by $\vee$ (or). For example, taking the domain of x to be the set of all Englishmen, the assertion that all Englishmen are either stupid or rogues does not imply that either all Englishmen are stupid or all Englishmen are rogues.

Q6. $(\exists x\, (P(x) \vee Q(x))) \leftrightarrow ((\exists x\, P(x)) \vee (\exists x\, Q(x)))$.

This is false if we replace $\vee$ (or) by $\wedge$ (and).

Mistakes often occur when more than one quantifier is applied to a proposition. We have the following rules.

Q7. $(\forall x \, \forall y \, P(x,y)) \leftrightarrow (\forall y \, \forall x \, P(x,y))$.

Q8. $(\exists x \, \exists y \, P(x,y)) \leftrightarrow (\exists y \, \exists x \, P(x,y))$.

Thus when the same type of quantifier is repeated the order does not matter, but note the following.

Q9. $(\exists x \, \forall y \, P(x,y)) \rightarrow (\forall y \, \exists x \, P(x,y))$.

In words this asserts: there exists a value of x for which $P(x,y)$ is true for all y implies that for all y there exists a value of x for which $P(x,y)$ is true. But **the opposite implication is generally false.**

For example, take the domain of the variables x and y to be the set of all natural numbers $\{0, 1, 2, 3, \ldots\ldots\}$. Take $P(x,y)$ to be the proposition $y \leqq x$. Then $\forall y \, \exists x \, (y \leqq x)$ (which reads: for all y, there exists a value of x for which $y \leqq x$) is true, because for any natural number y we can always find another natural number x which is greater than y. However $\exists x \, \forall y \, (y \leqq x)$ (which reads: there exists a value of x such that for all y we have $y \leqq x$) is false, because there is no natural number which is greater than all other natural numbers.

For a further discussion of quantifiers and logic in general we refer the interested reader to the very readable book: *Logic* by W. Hodges (Penguin Books 1977).

1.3 FORMAL PROOF

In mathematics those propositions which resemble the tautologies and the logically valid statements of logic are called **theorems**. If we want to give a **formal** proof of a theorem we must proceed as follows.

We must find a sequence of statements:

(1) A_1

(2) A_2

(3) A_3

. . . . . .

. . . . . .

(n) A_n

such that A_n is the required theorem and each A_i is either an axiom, a previously proved theorem, a tautology or logically valid statement, or follows from two earlier statements A_j and $A_j \to A_i$ by a process called **modus ponens**. This latter is the fundamental rule of classical logic. Thus we might have something like:

(1) A_1 axiom

(2) $A_1 \to A_2$ proven theorem

(3) A_2 from (1) and (2) by modus ponens

(4) $A_2 \to A_3$ proven theorem

(5) A_3 from (3) and (4) by modus ponens

(6) $A_3 \to A_4$ proven theorem

(7) A_4 from (5) and (6) by modus ponens

Thus A_4 is proved.

This would establish the theorem A_4 by a formal proof. Of course, in practice such a procedure is too cumbersome to give in detail, so we argue as in the two examples already given. However we should in principle be able to fill in a complete formal proof if challenged.

Note in particular two techniques used in informal argument, which we can now justify.

(1) Reductio ad absurdum. The formula $p \vee \sim p$ is a tautology as is easily checked by means of a truth table. This means that a statement is either true or false. (Please note that there are logics, such as intuitionistic logic discussed in 'Intuitionism' by Heyting, for which this is not true). To use this tautology we suppose that we have to prove the statement p. We suppose p is false and show that this leads to a contradiction. Since p is either false or true, it follows that p must be true. This was the method used in our two examples.

(2) Contra – positive proof. From the tautology $(p \to q) \leftrightarrow (\sim q \to \sim p)$, we conclude that if we want to prove that given p then q follows, then we can instead prove that given $\sim q$ then $\sim p$ follows.

1.4 GÖDEL'S RESULTS

In 1930 K. Gödel showed that if we try to prove theorems about the natural numbers from axioms then there will be some results which although true nevertheless cannot be proved by a chain of deduction of the kind we have discussed in section 1.3.

More precisely Gödel obtained a formula F stating some property of natural numbers, such that F is not formally deducible from the axioms and $\sim F$ is not formally deducible from the axioms. Intuition tells us that one of F or $\sim F$ should be true. In fact Gödel managed to show by an indirect argument that F is true.

Thus, although working axiomatically saves us from contradictions, the method may be unable to prove certain true results.

This may seem bad enough but an American logician Alonzo Church has proved that there exists a formula whose truth can neither be proved nor disproved by **any** argument used by present day mathematicians that can be formalised.

There have been various other attempts to provide mathematics with sound foundations e.g. intuitionism mentioned above, but all leave out some areas of mathematics. Recently Imre Lakatos has tried to escape from the strait jacket of the formal proof by looking more closely at the way in which mathematics actually is created. This has led him to look at proof not as a formal deduction but rather as a means of convincing others of the truth of given statements. The interested reader is referred to *Proofs and Refutations* by Lakatos published by Cambridge University Press.

1.5 SUMMARY

In practice what is an acceptable way of presenting mathematics in order to convince the reader of its truth is best discovered by actually looking at presentations known to be currently acceptable. For the student at whom this book is aimed, the rest of the book serves as such a guide. Later, as an independent judgement is developed, the student may well not be satisfied with this presentation and may demand more precision or perhaps more economy and elegance. Maybe we are looking for something as elusive as literary style.

All I hope to have done in this chapter is to have brought to the notice of the student some of the difficulties of presenting mathematics. For a much more detailed discussion of the topics of this chapter the student is referred to: *Concepts of Modern Mathematics* by Ian Stewart (Penguin Books 1975) and *Foundations of Mathematics* by Stewart and Tall (Oxford University Press 1977).

EXERCISES

1. The following propositions are known to be true:

$p \vee q$, $\sim(p \vee r)$, $p \vee s$. Which of the following propositions must necessarily be true?

$p \wedge q$, q, $\sim q$, s, $\sim s$.

2. Show that $p^2 - 1$ must be divisible by 8, if p is an odd natural number.

3. By means of truth tables find which of the following are tautologies.

(a) $(p \to q) \to ((r \to p) \to (r \to q))$

(b) $(a \vee \sim a) \to (b \vee (a \wedge b))$

(c) $a \to (b \to a)$

(d) $(a \vee b) \wedge \sim a \wedge \sim b \wedge (b \vee \sim a)$

(e) $(a \to b) \leftrightarrow (\sim b \to \sim a)$

(f) $a \to \sim\sim a$

(g) $\sim\sim a \to a$

4. Give a proof that $\sqrt{2}$ is not a rational number. (A rational number is any number of the form p/q where p and $q \neq 0$ are integers).

5. Taking as **axioms** the following formulae:

(a) $A \to (B \to A)$

(b) $(A \to (B \to C)) \to ((A \to B) \to (A \to C))$

(c) $(\sim B \to \sim A) \to ((\sim B \to A) \to B)$,

where A, B, C stand for any meaningful formulae, give **formal** proofs of the following formulae:

(d) $p \to p$

(e) $(\sim p \to p) \to p$

6. Give a proof by contradiction of the statement:

$$x \text{ is odd} \rightarrow x^2 \text{ is odd } (x \in \mathbf{Z}).$$

7. Give a **contra-positive** proof of the statement:

if $x \in \mathbf{R}$ and $|x| < y$ for **all** positive real numbers y, then $x = 0$.

8. Assuming the theorem that the sum of the interior angles of a triangle is 180°, prove by contradiction (**reductio ad absurdum**) that, in Euclidean geometry, if a straight line L cuts two other straight lines M and N in a plane so that the **alternate** angles are equal, then the lines M and N do **not** intersect.

9. Express **formally** (i.e. using the symbols $\forall$, $\exists$, $\rightarrow$, and so on) the following statements:

(a) There exists a real number whose square is 2.

(b) For all integers n, n^2 is not negative.

(c) For all x, y, if x and y are rational numbers, then so is $x + y$.

10. By giving a **counter-example** (i.e. an example for which the statement is false) show that the following conjecture is false.

Every equation of the form $a_0 + a_1 x + a_2 x^2 + \ldots + a_n x^n = 0$ with **rational** coefficients $a_0, a_1, \ldots\ldots, a_n$ has at least one rational solution.

11. Which of the following statements are correct?

(a) A theorem may be proved by drawing an accurate diagram where appropriate.

(b) It is possible to program a computer to prove or disprove **every** statement in mathematics.

(c) The negation of: 'All triangles are equilateral', is the statement: 'No triangles are equilateral'.

(d) The negation of: 'Some triangles are isosceles', is the statement: 'No triangles are isosceles'.

CHAPTER 2

Sets, Functions and Relations

2.1 SETS

Modern mathematics is based on the concept of the set. Since it is basic it cannot be defined but merely illustrated by pointing to examples.

A cricket team is a set whose members are the players in the team. The totality of all integers forms a set **Z** whose members are the individual integers.

The former is an example of a finite set since it has a finite number of members, while the latter is an infinite set.

As long as we work with finite sets or infinite sets like the integers, we are likely to remain free from trouble with the naive approach which is followed in this book. However, if we want to talk about very large sets, we may become involved in contradictions unless we are very careful. We have already considered in chapter 1 the Russell paradox. This arises from consideration of a very large 'set'; namely the Russell set which we write in symbols as: $S = \{A : A \notin A\}$. This means: all sets A such that A is not a member of A. As noted before we have the paradox: $S \in S \rightarrow S \notin S$ and $S \notin S \rightarrow S \in S$.

It is in order to avoid paradoxes like this that we resort to axiomatic set theory which is designed to prevent such sets arising. Alternatively we alter the logic. For example if $a \vee \sim a$ is not a tautology then it is possible for a statement to be both not true and not false. Thus there is another possibility besides $S \in S$ and $S \notin S$. This approach is followed by the intuitionists mentioned in chapter 1.

For an elegant discussion of axiomatic set theory see 'Naive Set Theory' by Paul Halmos (Springer). Although this book uses the word 'naive' in its title, it is in fact a very clear outline of how set theory can be developed axiomatically. In comparison our treatment really is naive.

2.2 SUBSETS

We say that S is a subset of T, written $S \subset T$ if and only if $S = \emptyset$, the empty set with no members, or $x \in S \rightarrow x \in T$.

In particular $T \subset T$. If $S \subset T$ but $S \neq T$, then S is said to be a **proper** subset of T.

One way of obtaining a set which is **allowable**, i.e. will not lead to contradictions, is to consider a subset of a set which is known already to be allowable.

This is usually done by 'picking out' from T the members of the subset S by some property $P(x)$ which defines them. In symbols we have: $S = \{x \in T : P(x)\}$. This reads: S is the set of all those members of T that have the property $P(x)$.

For example: $S = \{x \in \mathbf{Z} : x = 2y \text{ for some } y \in \mathbf{Z}\}$

This 'picks out' or defines the set S of even integers: $S = \{\ldots\ldots -4, -2, 0, 2, 4, 6, \ldots\ldots\}$.

2.3 THE POWER SET

The set of all subsets S of a given set T is denoted by $\mathcal{P}(T)$ and is called the power set of T.

For example: if $T = \{1, 2, 3\}$ then $\mathcal{P}(T)$ is the set:

$\{\phi, \{1\}, \{2\}, \{3\}, \{1, 2\}, \{1, 3\}, \{2, 3\}, \{1, 2, 3\}\}$

Note that $\mathcal{P}(\{1, 2, 3\})$ has 8 members. In general if T has n members then $\mathcal{P}(T)$ has 2^n members. This can be seen as follows. Number the n members of T with the natural numbers $1, 2, \ldots\ldots, n$. This means that we can take T to be the set $\{1, 2, 3, \ldots\ldots, n\}$. In obtaining a subset of T we can either include 1 or leave it out. With each of these two ways of dealing with 1 we can associate two ways of dealing with 2 and so on. Altogether there are $2 \times 2 \times 2 \times \ldots\ldots \times 2$ ways of constructing subsets of T, where there are n factors in the above product.

2.4 INTERSECTION AND UNION

From now on all sets that we consider will be understood to be subsets of one other set denoted by U. This is sometimes called the **universal set**. It remains fixed for a given problem but may change from problem to problem. In this way we always work with allowable sets as mentioned in section 2.2. This approach enables us to escape from having to give a full axiomatic development of set theory yet provides a reasonably satisfactory base on which to build the rest of the work in this book.

Given two sets A and B, we define the intersection of A and B written $A \cap B$ by:

$$A \cap B = \{x \in U : x \in A \text{ and } x \in B\}.$$

Thus $A \cap B$ consists of all members of A that are also members of B.

For example: if $A = \{1, 3, 5, 6, 8, 9\}$ and $B = \{2, 4, 6, 9, 11, 15\}$

then $A \cap B = \{6, 9\}$.

The union of the two sets A and B is defined by:

$$A \cup B = \{x \in U : x \in A \text{ or } x \in B\}.$$

Thus $A \cup B$ consists of all the members of A together with all the members of B.

For example: if $A = \{3, 7, 11\}$ and $B = \{2, 3, 6, 7\}$

then $A \cup B = \{2, 3, 6, 7, 11\}$. (In the above take $U = \mathbf{Z}$).

These ideas extend readily to more than two sets.
Thus if we have the sets $A_1, A_2, \ldots\ldots\ldots, A_n$, then

$$A_1 \cap A_2 \cap A_3 \cap \ldots\ldots \cap A_n = \{x \in U : x \in A_i \text{ for } \mathbf{all}\ i = 1, 2, \ldots\ldots, n\}$$

and $A_1 \cup A_2 \cup A_3 \cup \ldots\ldots \cup A_n = \{x \in U : x \in A_i \text{ for } \mathbf{some}\ i = 1, \ldots\ldots, n\}$

2.5 INDEX SET

It is convenient to write $A_1 \cap A_2 \cap A_3 \cap \ldots \cap A_n$ as

$\bigcap_{i \in \Delta} A_i$, and $A_1 \cup A_2 \cup A_3 \cup \cdots \cup A_n$ as $\bigcup_{i \in \Delta} A_i$, where

Δ is the set $\{1, 2, 3, 4, \ldots\ldots, n\}$.

A set Δ used in this way is called an **index** set.
This notation enables us to handle an infinite number of subsets as we can use any set for our index set Δ.

In general we have:

$$\bigcap_{i \in \Delta} A_i = \{x \in U : x \in A_i \text{ for } \mathbf{all}\ i \in \Delta\} \text{ and}$$

$$\bigcup_{i \in \Delta} A_i = \{x \in U : x \in A_i \text{ for } \mathbf{some}\ i \in \Delta\},$$

where Δ is an index set.

2.6 VENN DIAGRAMS

These afford a visual representation of some of the above ideas. We represent U by the area or points inside a rectangle, say, and we represent each subset by the area or points inside a simple closed curve, like a circle, contained in the rectangle.

Figure 2.6.1 shows the sets A, B, and C in this way.

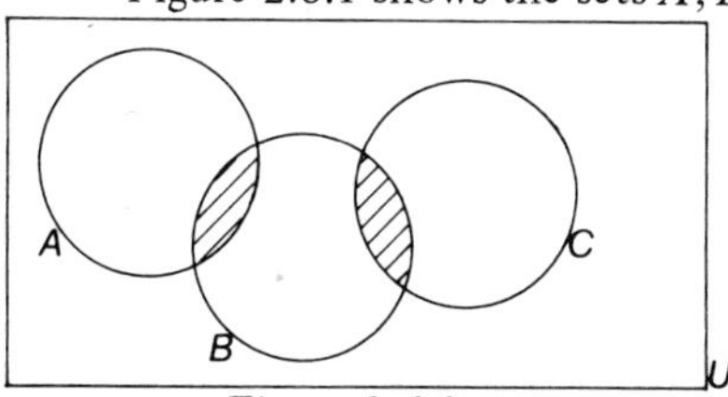

Figure 2.6.1

The shaded areas represent

$A \cap B$ and $B \cap C$.

2.7 RULES FOR MANIPULATING ∩ AND ∪

Idempotent rules:

(1) $A \cap A = A$

(2) $A \cup A = A$

Associative rules:

(3) $A \cap (B \cap C) = (A \cap B) \cap C$

(4) $A \cup (B \cup C) = (A \cup B) \cup C$

Distributive rules:

(5) $A \cap (B \cup C) = (A \cap B) \cup (A \cap C)$

(6) $A \cup (B \cap C) = (A \cup B) \cap (A \cup C)$

These equations may be shown to hold by use of the Venn diagrams. For example Figure 2.7.1 demonstrates equation (6). However this does not constitute a proof.

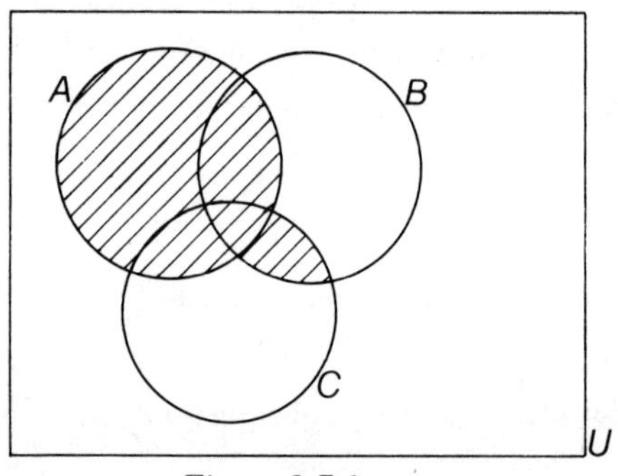

Figure 2.7.1

The standard way of proving that two sets S and T are equal is to show that $S \subset T$ and $T \subset S$. This can be done by showing that (a) $x \in S \rightarrow x \in T$ and

(b) $x \in T \rightarrow x \in S$.

We prove equation (5) and leave the reader to prove the other equations.

(5) $A \cap (B \cup C) = (A \cap B) \cup (A \cap C)$.

PROOF Let $x \in (A \cap B) \cup (A \cap C)$. Then

$$x \in (A \cap B) \text{ or } x \in (A \cap C).$$

Thus $x \in A$ and ($x \in B$ or $x \in C$). This means that

$$x \in A \cap (B \cup C).$$

Thus $(A \cap B) \cup (A \cap C) \subset A \cap (B \cup C)$. (a)

Let $x \in A \cap (B \cup C)$. Then

$x \in A$ and ($x \in B$ or $x \in C$). Hence

$x \in A$ and $x \in B$ or $x \in A$ and $x \in C$.

Thus $x \in (A \cap B) \cup (A \cap C)$.

Hence $A \cap (B \cup C) \subset (A \cap B) \cup (A \cap C)$ (b)

From our two inclusions (a) and (b) the equation (5) follows.

2.8 COMPLEMENTS

Relative to the universe U, the complement of the set A, denoted by A', is defined by $A' = \{x \in U : x \notin A\}$.
The relative complement $A - B$ means the set of all elements which are in A but are not in B. In particular $A' = U - A$. The following rules hold:

(1) $(A')' = A$

(2) $(A \cap B)' = (A' \cup B')$

(3) $(A \cup B)' = (A' \cap B')$

The proofs of these are left to the reader.

2.9 FUNCTIONS

Functions of a real variable were met with at school. For example: $x^2 + x + 1$ is a function of the real number represented by x. We write $f(x) = x^2 + x + 1$ and we have a graphical representation:

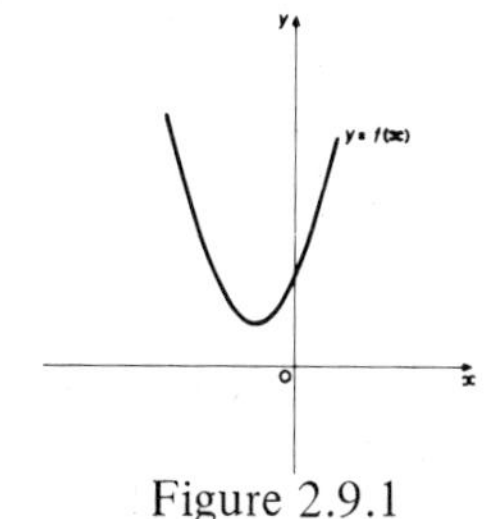

Figure 2.9.1

Another example is : $f(x) = \sin x$ with graph given by Figure 2.9.2

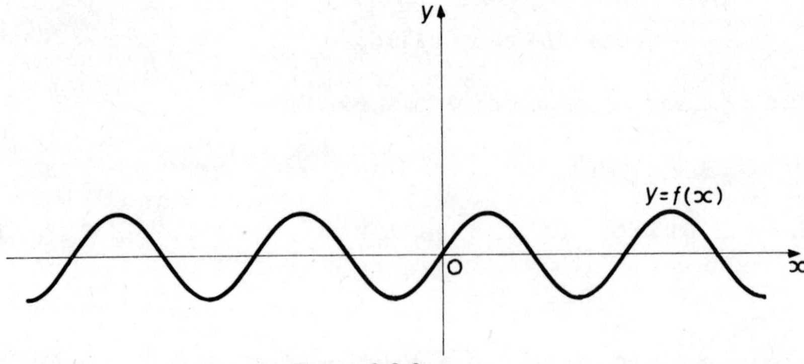

Figure 2.9.2

Both the above examples have the property that for a given $x \in \mathbf{R}$, where $\mathbf{R}$ denotes the set of real numbers, there is one and only one value of $f(x)$. Such a function is said to be **single-valued**.

A function such as $f(x) = \sin^{-1} x$, with the graph given in Figure 2.9.3, is called a **many-valued** function. However, we usually make it single-valued by restricting $f(x)$ to the so-called principal values.

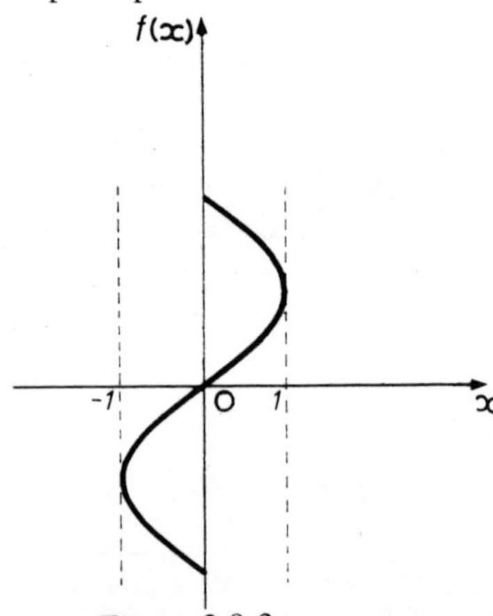

Figure 2.9.3

We shall be concerned only with single-valued functions.

Much of modern mathematics arises from the extension to a more general setting of concepts first associated with numbers. This process enables mathematics to be applied ever more widely. In this spirit we would like to extend the concept of function so that it applies when the variable is not necessarily a number.

To do this it is helpful to look at a function in a slightly different way from that outlined above. Instead of the usual two mutually perpendicular axes to represent x and $f(x)$, let us represent the set of real numbers, from which the values of the variable x are chosen, by certain points in the disc 1 (as in Venn diagrams), and the set of real numbers in which the values of $f(x)$ lie by certain points in the disc 2. For our purpose, the precise way in which this is done need not concern us. Thus we have Figure 2.9.4.

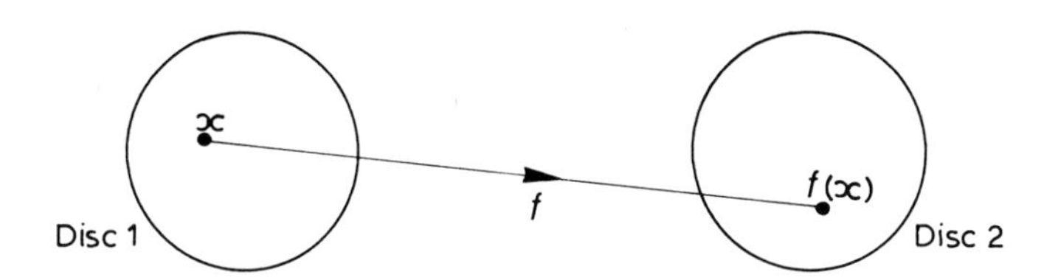

Figure 2.9.4

Then the function, say $f(x) = x^2 + x + 1$, can be thought of as a rule which tells us which unique real number $x^2 + x + 1$ in disc 2 is associated with a particular real number x in disc 1. Put in this form there is a natural generalisation to arbitrary sets which we give below.

DEFINITION 2.9.1 A function f from a set S to a set T is a rule which assigns to each member x of S a unique member y of T. We write $y = f(x)$ and call y the image of x under f. S is called the domain and T the codomain of f. The set of all members y of T that are images of members x of S is denoted by $f(S)$ and is called the image of f, or Im f. Thus Im $f = f(S) = \{y \in T \ : \ y = f(x)$ for **some** $x \in S\}$.

Note. There are several other terms in use for the concepts defined above. For example the term **range** is often used either in place of our term Im f or in place of codomain.

In terms of diagrams we have:

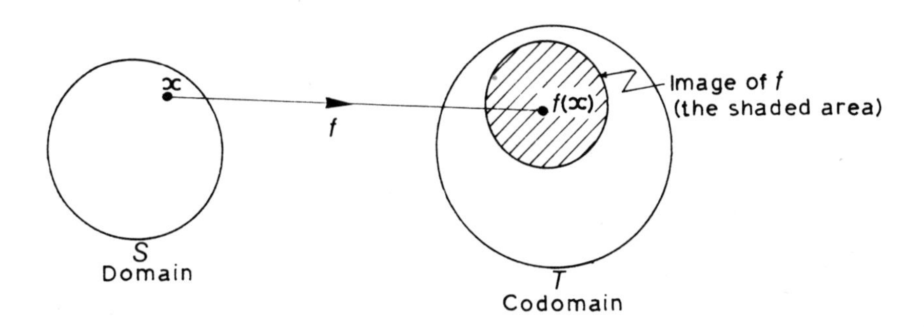

Figure 2.9.5

We sometimes write a function in the form:

$$f: x \longmapsto f(x), x \in S; \text{ or } S \xrightarrow{f} T.$$

EXAMPLE 2.9.1 A non-numerical example of the above is provided by taking

S = {all people in England}

T = {all the days in a leap year}, and

$f(x)$ = the birthday of x.

Instead of the word function we often use the word map or mapping, and talk about x mapping onto $f(x)$, or x being mapped onto $f(x)$.

In many cases the two sets S and T are the same set. This is so in the case of the function $f(x) = x^2 + x + 1$, $x \in \mathbf{R}$, discussed at the beginning of this section, provided we take the codomain T to be $\mathbf{R}$. Note that Im $f = \{x \in \mathbf{R}: x \geqq \frac{3}{4}\}$.

2.10 SPECIAL FUNCTIONS

Given $f: S \to T$, and $x_1, x_2 \in S$; if $f(x_1) = f(x_2)$ implies that $x_1 = x_2$, then we say that f is **one-one** or **injective**. In this case one and only one $x \in S$ maps onto a given $y \in T$.

If for any given $y \in T$ there always exists $x \in S$ such that $y = f(x)$, then f is said to be **onto** or **surjective**. This means that $T = f(S)$; that is the image of f is the whole of T. Figure 2.10.1 illustrates this idea.

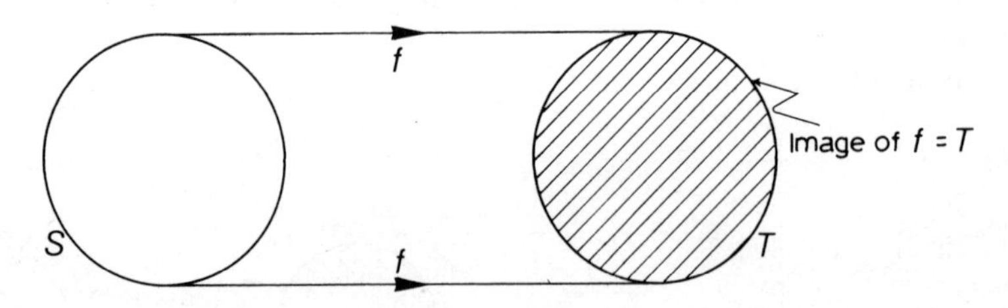

Figure 2.10.1

A function which is both one-one and onto is said to be a **one-one correspondence** or a **bijection** or to be **bijective**. When $S = T$ a bijective function is called a **permutation** of S.

The function i_S defined by $i_S(s) = s$, for all $s \in S$, is called the **identity** function on S.

2.11 COMPOSITION OF FUNCTIONS

If we have two functions $S \xrightarrow{f} T$ and $T \xrightarrow{g} W$, where S, T and W are sets, then by following f with g we obtain a function $S \xrightarrow{h} W$ called the composition of f and g. We write : $h = g \circ f$ and define h formally by :

$$h(s) = (g \circ f)(s) = g(f(s)), \text{ where } s \in S.$$

We have: $s \longmapsto f(s) \longmapsto g(f(s))$.

EXAMPLE 2.11.1

Take $S = T = W = \mathbf{R}$.

Let $f : x \longmapsto x^2, x \in \mathbf{R}$ and $g : x \longmapsto 3x + 6, x \in \mathbf{R}$.

Then $g \circ f : x \longmapsto 3x^2 + 6, x \in \mathbf{R}$.

At school you probably called this a 'function of a function'.
By considering the sequence of functions:

$$S \xrightarrow{f} T \xrightarrow{g} W \xrightarrow{h} U$$

it is easy to check that the associative rule holds for the composition of functions i.e. $h \circ (g \circ f) = (h \circ g) \circ f$ or, in a form suitable for checking,

$$h((g \circ f)(s)) = (h \circ g)(f(s)), \text{ where } s \in S.$$

In fact both sides of the last equation are equal to $h(g(f(s)))$.

2.12 INVERSES

Given a function $S \xrightarrow{f} T$, we say that a function $T \xrightarrow{g} S$ such that $g \circ f = i_S$ is a **left inverse** of f. Similarly we say that f in the above is a **right inverse** of g.

In terms of left and right inverses there are useful characterisations of one-one and onto functions.

We state these as:

Theorem 2.12.1

(a) f is one-one if and only if f has a left inverse.

(b) f is onto if and only if f has a right inverse.

Proof

(a) Let f be one-one and define $g : T \longrightarrow S$ by $g(t) = s$, if $t \in \operatorname{Im} f$, where s is the unique $s \in S$ such that $f(s) = t$, and $g(t) = x$ if $t \notin \operatorname{Im} f$, where x is any one member of S.

We have : $(g \circ f)(s) = s$ for all $s \in S$.

Hence $g \circ f = i_S$ as required.

Now let f have a left inverse g and suppose that

$$f(s_1) = f(s_2). \text{ Then } g(f(s_1)) = g(f(s_2)).$$

Thus $(g \circ f)(s_1) = (g \circ f)(s_2)$. But $g \circ f = i_S$.

Hence $i_S(s_1) = i_S(s_2)$. Finally $s_1 = s_2$.

Thus $f(s_1) = f(s_2)$ implies that $s_1 = s_2$. By definition f is one-one, as required.

(b) Let f be onto. Define g: $T \longrightarrow S$ by $g(t) = s$ where $t \in T$ and s is any member of S such that $f(s) = t$. We know there must be at least one such s because f is onto. (In the case of infinite sets S and T this selection of a suitable s may involve the **'axiom of choice'**. For discussion of this notion see page 256 '*Foundations of Mathematics*' by Stewart and Tall, or '*Naive Set Theory*' by Halmos).

Now we have: $(f \circ g)(t) = f(g(t)) = f(s) = t$, for all $t \in T$.

Thus $f \circ g = i_T$ as required.

Conversely let f have a right inverse g so that $f \circ g = i_T$.

Let t be any member of T.

Then $t = i_T(t) = (f \circ g)(t) = f(g(t)) = f(s)$, where $s = g(t)$.

Thus $f(s) = t$.

Hence f is onto.

Suppose that f is both one-one and onto. By the above theorem it follows that f has a left inverse g: $T \longrightarrow S$ and a right inverse h: $T \longrightarrow S$. We have: $g \circ f = i_S$ and $f \circ h = i_T$.

However $h = i_S \circ h = (g \circ f) \circ h = g \circ (f \circ h) = g \circ i_T = g$.

Thus $f \circ g = i_T$ and $g \circ f = i_S$.

Such a g is called a two sided inverse of f, or simply an **inverse** of f. But we can say more.

Let g^1 be **another** inverse of f. Then

$$g = g \circ i_T = g \circ (f \circ g^1) = (g \circ f) \circ g^1 = i_S \circ g^1 = g^1.$$

Hence the inverse g of f is **unique**.

Conversely if f has an inverse then this inverse is both a right inverse and a left inverse. Hence by theorem 2.12.1 f is one-one and onto.

These results can be summed up conveniently as:

THEOREM 2.12.2

If f has an inverse then f is one-one and onto. If f is one-one and onto then f has an inverse which is **unique**. It is usual to denote this **unique** inverse by f^{-1}. It is characterised by: $f^{-1}(y) = x$ if and only if $y = f(x)$, where $x \in S$ and $y \in T$, and f: $S \to T$.

NOTE however that in set theory, **even when f is not one-one**, the symbol f^{-1} is often used as defined by the following.

Let $f : S \longrightarrow T$. Let $W \subset T$.

Then $f^{-1}(y) = \{x \in S : f(x) = y\}, y \in T$, and $f^{-1}(W) = \{x : f(x) \in W\}$.

In other words $f^{-1}(y)$ denotes the set of all those elements which f maps onto y. Similarly $f^{-1}(W)$ denotes the set of all those elements which f maps into the set W.

2.13 RELATIONS

In the everyday world we are all familiar with the idea of relations: brothers, sisters, aunts, cousins. We are also familiar with the idea of comparing numbers, particularly when they represent sums of money. We would like to capture the essence of this idea of 'relating' or 'comparing' two objects.

When we relate or compare objects in some population we are in effect picking out from the population, or set of objects, pairs of these objects, in fact ordered pairs, namely the pairs being related.

For example, if the relation being considered is that of father and son, and the population is the set of people in a full coach on a holiday trip, then the pairs selected will each consist of a father and son.

If the set in which we are interested is finite this suggests labelling our objects 1, 2, 3, 4, 5, ..., n and describing our relation by listing the ordered pairs that exhibit the relation concerned. This may result in the pairs (1, 3), (2, 4), (5, 1) being selected.

For example, suppose we have a football team labelled 1, 2, 3, ..., 11, and the relation connecting players x and y is that they share a flat. We could describe this relation by a list of ordered pairs, say, (2, 5), (5, 2), (11, 1), (1, 11). This means that players 2 and 5 share a flat as do players 1 and 11.

This suggests how we might define the concept of a relation on an abstract set S.

First we define the set $S \times S$, called the **Cartesian product** of the set S with itself, to be the set of all ordered pairs (a,b), where a and b are members of S. For example if $S = \{1, 2, 3\}$ then

$$S \times S = \{(1, 1), (1, 2), (1, 3), (2, 1), (2, 2), (2, 3), (3, 1), (3, 2), (3, 3)\}.$$

The formal definition of relation is given by

DEFINITION 2.13.1 A **relation** R on a set S is defined to be a **subset** of the set $S \times S$.

Thus R is a relation on S if and only if $R \subset S \times S$.

By an abuse of our notation, we often write xRy to mean that x is related to y by the relation R instead of writing $(x, y) \in R$.

Note that if S is a finite set with n members we can always replace S by the set $\{1, 2, 3, 4, \ldots\ldots, n\}$ and represent R by a set of ordered pairs of the form (2, 3), (5, 1), etc.

Another way of representing the relation R is by a graph, as follows.

The points of the graph represent the members of S and an arrow connects x to y if and only if $(x, y) \in R$. Thus if $S = \{1, 2, 3\}$ and $R = \{(1, 1), (2, 3), (3, 1)\}$, we have the graph shown in Figure 2.13.1.

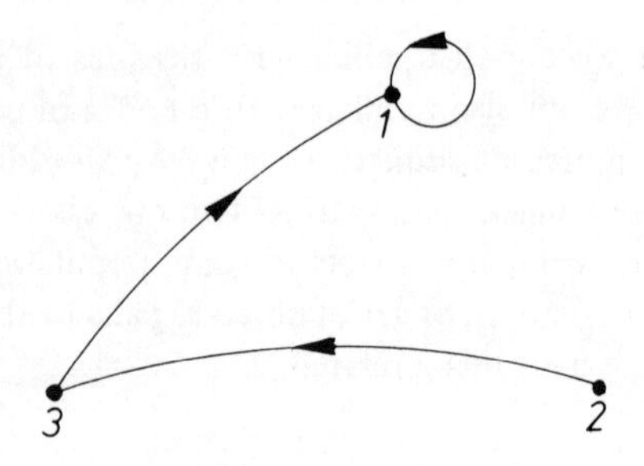

Figure 2.13.1

2.14 EQUIVALENCE RELATIONS

One of the major activities in any science is the classification of the objects of study. This amounts to a partition of the set of objects into mutually disjoint subsets. Taking S as the set of objects and S_i as a typical subset of the partition, we have

$$S = \bigcup_{i \in \Delta} S_i, \text{ where } \Delta \text{ is some index set, and}$$

$$S_i \cap S_j = \phi \text{ if } i \neq j.$$

Figure 2.14.1 illustrates this idea.

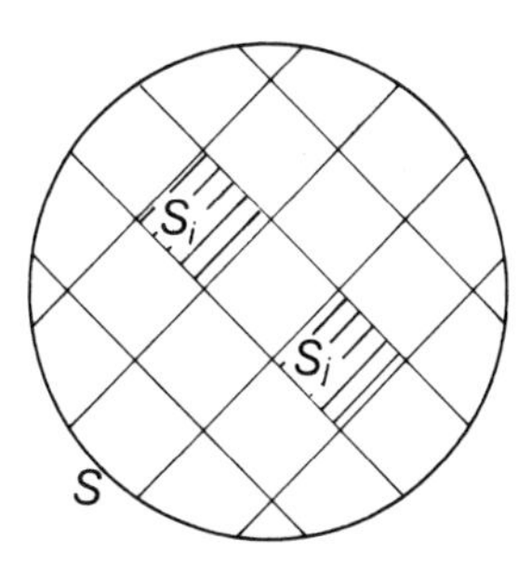

Figure 2.14.1

Let us now define a relation on S by use of the partition. We define R on S by:

$$xRy \leftrightarrow x, y \in S_i \text{ for some } i \in \Delta.$$

In other words x and y are related if and only if they are members of the **same** subset S of the partition.

We note the following properties of R.

(**E**1) xRx (reflexive).

(**E**2) If xRy and yRz, then xRz (transitive).

(**E**3) If xRy, then yRx (symmetric).

Here x, y and z are arbitrary members of S.

As indicated, these are called the **reflexive**, **transitive**, and **symmetric** properties of R, respectively.

Now let us suppose that we have a relation R on S which has these three properties. Let S_x be the subset of S defined by:

$$y \in S_x \leftrightarrow xRy, \text{ or } S_x = \{y \in S : xRy\}.$$

Then by (1) $x \in S_x$. Thus $S =: \bigcup_{x \in S} S_x$, using S itself as an index set.

Now suppose $y \in S_x \cap S_z$. Then xRy and zRy.

By (2) and (3) we have xRz. If $u \in S_z$ then zRu.

By (2) xRu. Thus $u \in S_x$. Hence $S_z \subset S_x$.

A similar argument, or the symmetry of the expression $y \in S_x \cap S_z$ in x and z, yields $S_x \subset S_z$.

Altogether we conclude that $S_x = S_z$.

Thus either $S_x \cap S_z = \phi$ or $S_x = S_z$.

Discarding repeated subsets in $\bigcup_{x \in S} S_x$, it follows that

$$S = \bigcup_{i \in \Delta} S_{x_i} \text{ and } S_{x_i} \cap S_{x_j} = \phi \text{ , if } x_i \neq x_j$$

for suitable $x_i \in S$.

Thus the relation R has partitioned S into subsets.

Any relation R which has the three properties (**E**1), (**E**2), (**E**3) is called an **equivalence relation** and the subsets of the resulting partition are called **equivalence classes.**

EXAMPLE 2.14.1 An important example of an equivalence relation is as follows. Take $S = \mathbf{Z}$ (the set of integers) and take a fixed integer m. Define R by:

xRy if and only if $m \mid (x-y)$, where $\mid$ denotes 'divides'.

The reader is invited to carry out the routine check that this does define an equivalence relation on **Z**. The equivalence classes are usually denoted by: $\overline{0}, \overline{1}, \overline{2}, \overline{3}, \ldots\ldots, \overline{(m-1)}$, where $\overline{r}$ is the equivalence class containing the integer r.

Note that there are m distinct classes corresponding to the m possible positive remainders on dividing an integer by m. For this reason the equivalence classes are often called **residue classes modulo** m in this particular case.

An equivalence relation represented by a graph results in a graph of the following kind.

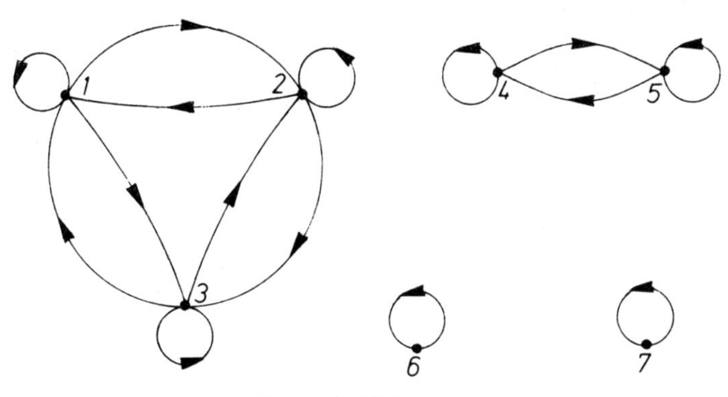

Figure 2.14.2

This is the relation R on $S = \{1, 2, 3, 4, 5,,6, 7\}$

given by $R = \{(1,1), (2,2), (3,3), (4,4), (5,5), (6,6), (7,7), (1,2), (2,1), (1,3), (3,1)$
$(2,3), (3,2), (4,5), (5,4)\}$.

Note that the graph consists of several distinct connected components, each component corresponding to an equivalence class. Each component has all possible arrows between the points in that component i.e. is a complete directed graph.

EXAMPLE 2.14.2 Another important example of an equivalence relation is as follows.

Let f be a function from a set S to a set T.

Define a relation R on S by: xRy if and only if $f(x) = f(y)$.

Thus members of S are related if and only if they are mapped by f onto the same member of T. This relation obviously partitions S as can be seen in Figure 2.14.3. Hence R is an equivalence relation.

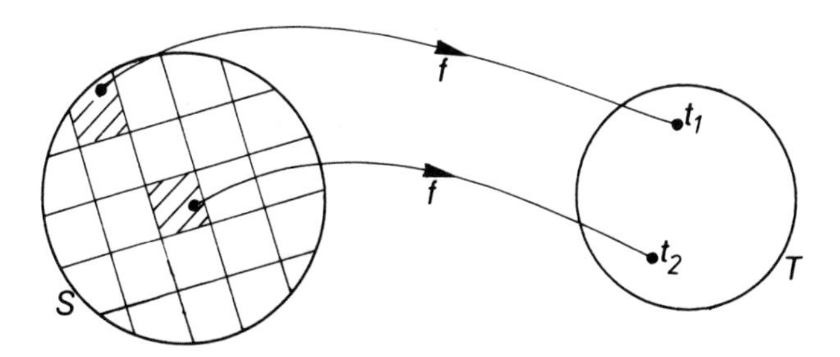

Figure 2.14.3

All members of a subset of S map onto the same member of T.

2.15 ORDER RELATIONS

Apart from classification, which led to the idea of an equivalence relation, another major activity in everyday life is the placing of objects into hierarchies. If the set of objects happens to be the integers we have the natural ordering:

$$\ldots\ldots\ldots -4, -3, -2, -1, 0, 1, 2, 3, 4, 5, \ldots\ldots\ldots$$

which we note has neither a beginning nor an end.

This ordering is an example of a relation with special properties.

Let us define R on $\mathbf{Z}$ by:

$$xRy \text{ if and only if } x \leqq y,$$

that is $x = y$ or x comes before y in the natural ordering above, reading from left to right. This relation on $\mathbf{Z}$ has the following properties.

(1) xRx (reflexive).

(2) If xRy and yRz then xRz (transitive).

(3) If xRy and yRx then $x = y$ (anti-symmetric).

Here x, y and z are members of $\mathbf{Z}$.

Note that this differs from an equivalence relation in one respect i.e. anti-symmetry instead of symmetry. In an order relation if we have xRy we must never have yRx unless $x = y$.

In the graphical representation this means that if an arrow connects two points there must be no return arrow. The system is one-way.

Note also that anti-symmetry does not exclude symmetry. A relation can be symmetric and anti-symmetric at the same time.

For example take $S = \{1, 2, 3\}$ and

$R = \{(1,1), (2,2), (3,3)\}$ with the graph in Figure 2.15.1

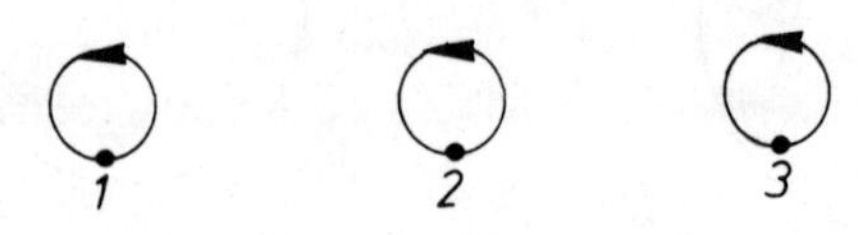

Figure 2.15.1

In this case R is both an equivalence relation and an order relation.

Formally we say that a relation R on a set S is an **order relation** if and only if it is reflexive, transitive, and anti-symmetric.

The natural order on $\mathbf{Z}$ is more than an order relation in that if x and y are two integers then either $x \leqq y$ or $y \leqq x$ for all $x, y \in \mathbf{Z}$.
This is not necessary for the order relation on S defined above. For that reason the latter is sometimes called a **partial** order. However, we shall distinguish between the two order relations by calling the order relation, in which either xRy or yRx for all $x, y \in S$, a **linear** or **complete** ordering.

EXAMPLE 2.15.1

Let $S = \{1, 2, 3\}$

Let $R = \{(1,1), (2,2), (3,3), (2,3)\}$

Then R is an order relation on S, but is not a complete order relation. Its graph is given in Figure 2.15.2

Figure 2.15.2

2.16 THE CARDINAL NUMBER OF A SET

The definition and theory of cardinal and ordinal numbers lies beyond the scope of this book. We refer the interested reader to 'Naive Set Theory' by Halmos. An elementary and readable discussion of cardinal numbers is contained in 'The Foundations of Mathematics' by Stewart and Tall. Here we shall be content to give the bare minimum required in the rest of the book.

For every set A there exists a set called the cardinal number of A. It is denoted by $|A|$.

Let A and B be any two sets.

Then the cardinal number has the characteristic property that:

$|A| = |B|$ if and only if there exists a one-one correspondence

(bijection) $f : A \to B$.

In the case of a finite set with n members the cardinality is the set $\{0, 1, 2, 3, 4, \ldots\ldots, n-1\}$.
In fact this is just the number n if we use the following method of **defining** the natural numbers in terms of sets rather than regard them as being intuitively known.

We **define** 0 to be the empty set ϕ. Then in succession we **define**:

$1 = \{\phi\} = \{0\}$, the set whose only member is 0,

$2 = \{\phi, \{\phi\}\} = \{0,1\}$, the set with just '2' members 0 and 1,

$3 = \{\phi, \{\phi\}, \{\phi, \{\phi\}\}\} = \{0,1,2\}$, the set with just the '3' members

0, 1, 2, and so on; eventually:

$n = \{0,1,2,3,4, \ldots\ldots, n-1\}$, the set whose members are those

natural numbers **already defined.**

Thus the cardinal number is a way of generalising to infinite sets the idea of the number of elements in the set.

The usual arithmetic operations such as addition, multiplication, exponentiation, can be extended to cardinal numbers. For example, addition and multiplication can be defined as follows.

Let n and m be cardinal numbers.

Take two **disjoint** sets A and B such that

$$|A| = n \text{ and } |B| = m.$$

Define:

$$n + m = |A \cup B| \quad \text{and} \quad n \times m = |A \times B|,$$

where $A \times B$ is the Cartesian product of the sets A and B.

However a rather strange arithmetic results.

For example, if

$$\mathbf{N} = \{0,1,2,3,4 \ldots\ldots, \ldots\}$$

is the set of all the natural numbers including 0, and the Hebrew aleph $\aleph_0$ denotes $|\mathbf{N}|$, then we have:

$$\aleph_0 + \aleph_0 = \aleph_0, \quad \aleph_0 \times \aleph_0 = \aleph_0.$$

It is usual to call a set S which is finite **or** has cardinality $\aleph_0$, that is $|S| = \aleph_0$, a **countable** or **enumerable** set. A set S with cardinality $\aleph_0$ is usually said to be **denumerable** or **countably infinite**. However usage varies here.

We prove now two basic results which we state as:

THEOREM 2.16.1

(a) Let **Q** denote the set of rational numbers. Then

$$|\mathbf{Q}| = \aleph_0.$$

(b) Let **R** denote the set of real numbers. Then

$$|\mathbf{R}| \neq \aleph_0.$$

(We usually denote $|\mathbf{R}|$ by the symbol c).

PROOF

(a) We arrange the positive rationals in an array as follows:

$1/1$	→	$1/2$		$1/3$	→	$1/4$	. . . .
	↙		↗		↙		
$2/1$		$2/2$		$2/3$		$2/4$	. . . .
↓	↗		↙				
$3/1$		$3/2$		$3/3$		$3/4$	. . . .
	↙						
$4/1$		$4/2$		$4/3$		$4/4$	. . . .

The rational number p/q, where p and q are positive integers, occurs in the pth row and qth column of the array. We count the rationals in the order shown by the arrows. There will be repetitions.

For example $1/2$ occurs as $2/4$, $3/6$, $4/8$ etc.

We strike out repetitions. We order the negative rationals in the same way. Finally we order the set of all the rationals **Q** by first taking 0 then alternating between the sets of positive and negative rationals.
This gives a list starting off as follows:

$$0, {}^1/_1, -{}^1/_1, {}^1/_2, -{}^1/_2, {}^2/_1, -{}^2/_1, {}^3/_1, -{}^3/_1, {}^1/_3, -{}^1/_3, \ldots\ldots\ldots$$

It is possible to find a formula which gives the *n*th rational in this list. We leave this to the reader.

The above process gives a one-one correspondence between **N** and **Q**.

Thus $|\mathbf{Q}| = \aleph_0$.

(b) To represent the members of **R** we use the usual decimal notation. To avoid ambiguity decimals ending like $\cdots\cdot 9999\ldots$ will be written in the form $\cdots\cdot 10000\ldots$

Suppose $|\mathbf{R}| = \aleph_0$. This means that we can list the real numbers as follows.

$$n_1 \cdot a_1^1\ a_2^1\ a_3^1 \ldots\ldots\ldots\ldots a_r^1 \ldots\ldots\ldots\ldots$$

$$n_2 \cdot a_1^2\ a_2^2\ a_3^2 \ldots\ldots\ldots\ldots a_r^2 \ldots\ldots\ldots\ldots$$

$$\vdots$$

$$n_i \cdot a_1^i\ a_2^i\ a_3^i \ldots\ldots\ldots\ldots a_r^i \ldots\ldots\ldots\ldots$$

$$\vdots$$

Here n_i is the integral part of the real number.

Now define a real number $k \cdot b_1\ b_2\ b_3 \ldots\ldots\ldots\ldots b_r \ldots\ldots\ldots\ldots,$

where $b_r = 3$ if $a_r^r \neq 3$, and $b_r = 2$ if $a_r^r = 3$; $r = 1,2,3,\ldots\ldots\ldots$

Since the real numbers can be listed, this real number that we have just defined must occur somewhere in the list.

Let it occur in the mth place. Then

$$k \cdot b_1\, b_2\, b_3 \ldots\ldots b_r \ldots = n_m \cdot a_1^m\, a_2^m \ldots\ldots a_r^m \ldots\ldots$$

However by construction $b_m \neq a_m^m$.

Thus $\quad k \cdot b_1\, b_2\, b_3 \ldots\ldots b_r \ldots . \neq n_m \cdot a_1^m\, a_2^m \ldots\ldots a_r^m \ldots\ldots$

From this contradiction we conclude that the real numbers cannot be listed.

Thus $\quad |\mathbf{R}| = c \neq \aleph_0$.

This concludes the proof of the theorem.

EXERCISES

1. Let S and T be sets.

Let $|S| = 9$, $|T| = 7$, and $|S \cap T| = 3$.

Find $|S \cup T|$.

2. Let S and T be finite sets.

Find an expression for $|S \cup T|$ in terms of $|S|$, $|T|$, and $|S \cap T|$.

3. In a survey of 100 students it was found that the numbers studying various languages were as follows: Spanish 28, German 30, French 42, Spanish and German 8, Spanish and French 10, German and French 5, all three languages 3.

Find (a) the number studying no language and

(b) the number studying French alone.

4. Let $A = \{1,2,3\}$ and let $f\colon A \to A$ be the function defined by the set of

ordered pairs $\{(1,2), (2,3), (3,1)\}$. This means that $(n, m) = (n, f(n))$.

Let g: $A \to A$ be defined by the set $\{(1,2), (2,1), (3,3)\}$.

Find $f \circ g$, $g \circ f$, $f \circ f$, $f \circ g \circ f$, and f^{-1}.

Verify that $f \circ g \circ f = g$, and that $f \circ f = f^{-1}$. (Here $f \circ g$ means first g then f).

5. (a) Let f: $A \to B$ and g: $B \to C$ both be surjective.

Prove that $g \circ f$: $A \to C$ is also surjective.

(b) Let f: $A \to B$ and g: $B \to C$ both be injective.

Prove that $g \circ f$: $A \to C$ is also injective.

6. If f: $A \to B$ and $C \subset A$, define $f(C)$ to be the set $\{f(x) \in B : x \in C\}$.

Two subsets S_1, S_2 of a set S are said to be **complementary** in S if $S_1 \cap S_2 = \phi$ and $S_1 \cup S_2 = S$.

Let f: $A \to B$ be injective and let A_1, A_2 be complementary in A.

Prove that $f(A_1)$ and $f(A_2)$ are complementary in $f(A)$.

7. Prove that the set $\{n \in \mathbf{N} : n \geqq k\}$, where **N** is the set of natural numbers, and k is a **fixed** natural number, has cardinality $\aleph_0$.

8. Prove that the set of real numbers $\{x \in \mathbf{R} : -1 < x < 1\}$ has cardinality c, the same as the set **R**.

9. Let S be any non-empty set. Let $\mathcal{P}(S)$ be the set of all subsets of S. Prove that $\mathcal{P}(S)$ and S do not have the same cardinality.

10. A relation R is defined on the set of positive integers $\mathbf{Z}^+$ by the rule aRb if and only if a and b have a common factor other than 1. Also $1R1$.

Prove that R is reflexive and symmetric, but is not anti-symmetric and is not transitive.

11. Considering only the reflexive, symmetric, and transitive properties of a relation, there are 8 different types of relation. (According to whether the relation has or has not the property concerned).

Give one example of each of the 8 types of relation on the set $S = \{1,2,3,4,5,6\}$.

Describe the relation both by specifying a subset of $S \times S$ and by giving a directed graph.

CHAPTER 3

Numbers and Induction

3.1 AXIOMS FOR RINGS

As briefly outlined in section 2.16 it is possible to construct the natural numbers, including zero, within the context of set theory. However, for simplicity, we shall adopt a naive approach. We shall assume both the natural numbers and the integers to be known. Sometimes it is convenient to include zero as a natural number and sometimes not. This will usually be clear from the context.

We shall also assume that the integers under addition and multiplication have certain properties. Other systems that we shall meet have similar properties. To refer to these easily without constant repetition it is convenient to have a universal description of all such systems. First we have a definition.

DEFINITION 3.1.1

Let S be a set. A **binary operation** on S is a function from the domain $S \times S$ into the codomain S.

EXAMPLE 3.1.1

Take $S = \mathbf{Z}$. Define f: $\mathbf{Z} \times \mathbf{Z} \to \mathbf{Z}$ by $f(n,m) = n + m$.

This exhibits the usual addition of integers as a binary operation.

If f is a binary operation on a set S it is the usual custom to denote $f(a,b)$ by ab or $a + b$, where a and b are arbitrary members of S.

We are now in a position to define various kinds of ring.

DEFINITION 3.1.2.

(a) A **ring** is a set S, together with a binary operation of addition + and a binary operation of multiplication $\cdot$ defined on S, such that the following axioms are satisfied:

(1) $(a + b) + c = a + (b + c)$ associative rule.

(2) There exists a **unique** element of S denoted by 0, the **zero** element, such that : $a + 0 = a = 0 + a$.

(3) To each a there exists a **unique** element of S, denoted by $(-a)$, called the **additive inverse** of a, such that: $a + (-a) = 0 = (-a) + a$.

(4) $a + b = b + a$ commutative rule.

(5) $a(bc) = (ab)c$ associative rule.
(Here we have dropped the $\cdot$ for multiplication, and symbolise multiplication by the juxtaposition of the symbols a,b,c, etc.).

(6) $a(b + c) = (ab) + (ac)$

(7) $(b + c)a = (ba) + (ca)$

} distributive rules.

Note that in all the above, a, b, c, etc. represent elements of S. These elements are arbitrary except in those cases stated in (2) and (3), namely the zero 0 and, for a given a, the inverse $(-a)$.

(b) **A ring with identity** is characterised by axioms

(1) to (7) together with the extra axiom:

(8) There exists a **unique** element e, such that :

$$ae = a = ea \text{ for all } a \in S.$$

We call e the **identity** of S.

(c) **A commutative ring** is characterised by axioms

(1) to (7) together with the extra axiom:

(9) $ab = ba$ for all $a, b \in S$.

(d) **A commutative ring with identity** is characterised by axioms (1) to (9).

(e) **An integral domain** is characterised by axioms (1) to (9) together with:

(10) If $ab = 0$ then either $a = 0$ or $b = 0$.

(f) **A non-commutative field, skew field,** or **division ring** is characterised by axioms (1) to (8) together with:

(11) If a is any **non-zero** element of S, then there exists a **unique** element of S, denoted by a^{-1} and called the **multiplicative inverse** of a, such that:

$$a\,a^{-1} = e = a^{-1}\,a, \text{ where } e \text{ is the identity of } S.$$

(g) **A field** is characterised by axioms (1) to (8) together with axioms (9) and (11).

We can now describe our assumptions about **Z**, the set of integers, by saying that **Z** is an integral domain under the usual binary operations of + and ×. The zero of axiom (2) is the number 0. The identity of axiom (8) is the number 1. -6 is the additive inverse of 6, and so on.

If we regard the rational numbers (i.e. numbers of the form $^p/_q$ where p and q are integers with $q \neq 0$) as intuitively known, then the set **Q** of all rational numbers with the usual binary operations of addition and multiplication is a field.

The multiplicative inverse of $^p/_q$, when $p \neq 0$, $q \neq 0$, is $^q/_p$.

$$\text{Let } \mathbf{Z}_m = \{\overline{0}, \overline{1}, \overline{2}, \ldots\ldots, \overline{(m-1)}\}$$

be the set of residue classes modulo m introduced in Example 2.14.1.

Define binary operations of addition and multiplication as follows:

$$\overline{a} + \overline{b} = \overline{(a+b)} \text{ and } \overline{a} \cdot \overline{b} = \overline{(a \cdot b)}.$$

Thus $\overline{3} + \overline{7} = \overline{10} = \overline{2}$, and $\overline{3} \cdot \overline{7} = \overline{(21)} = \overline{5}$, when $m = 8$.

We have to check that these definitions are valid in the sense that if

$$\overline{a} = \overline{a}',\ \overline{b} = \overline{b}', \text{ then } \overline{a} + \overline{b} = \overline{a}' + \overline{b}', \text{ and } \overline{a} \cdot \overline{b} = \overline{a}' \cdot \overline{b}'.$$

This point is taken up in Exercises 3.

If m is a product of factors, $m = rs$, $\mathbf{Z}_m$ is a ring with identity $\overline{1}$, but **not** an integral domain, since

$$\overline{0} = \overline{m} = \overline{rs} = \overline{r} \cdot \overline{s}, \text{ with } \overline{r} \neq \overline{0}, \overline{s} \neq \overline{0},$$

where $\overline{0}$ acts as the zero of the ring $\mathbf{Z}_m$. Axioms like (1) are verified as follows:

$$(\overline{a}+\overline{b}) + \overline{c} = (\overline{a+b}) + \overline{c} = (\overline{(a+b) + c}) = \overline{a+(b+c)} = \overline{a} + (\overline{b+c})$$

$$= \overline{a} + (\overline{b} + \overline{c}).$$

For further details see Exercises 3 and the solutions.

Further examples of rings of various kinds will arise later in the book, particularly when matrices become available.

3.2 THE PRINCIPLE OF INDUCTION

We shall assume that in their natural order: 1,2,3,4,5, ... , n, . . ., the positive integers (or natural numbers without the zero) are **well-ordered**. This means that any **non-empty** set of positive integers has a **first** or **least** member. Such a number n is characterised by the property that if x is any element in the set concerned then $n \leqq x$.

We state the **principle of induction** in two forms:

THEOREM 3.2.1

(a) Let $P(n)$ be a statement involving a positive integer n.

Let (1) $P(1)$ be true,

(2) $P(n) \rightarrow P(n+1)$.

Then $P(n)$ is true for all positive integers n.

(b) Let (1) $P(1)$ be true,

(2) $\{P(1), P(2), \ldots\ldots, P(n)\} \rightarrow P(n+1)$.

Then $P(n)$ is true for all positive integers n.

PROOF

Assume the well-ordering principle stated above. Suppose that $P(k)$ is false for some $k \in \mathbf{Z}^+$, the set of positive integers. Let K be the set of all such k. By the above supposition $K \neq \phi$.
By the well-ordering of $\mathbf{Z}^+$, K has a first member. Let this first member be k_0. Since $P(1)$ is true by the hypothesis of the theorem, we must have $k_o > 1$.

Thus $\{P(1), P(2), P(3), \ldots\ldots, P(k_o - 1)\}$ is true.

Hence by (b) 2, we conclude that $P(k_o)$ is true. This is a contradiction to the choice of k_o. We conclude that $P(k)$ cannot be false for any $k \in \mathbf{Z}^+$.

Thus $P(n)$ holds for all n. This concludes the proof of (b).

We now prove that (b) implies (a).
Assume the truth of (b) and suppose that (1) and (2) of (a) also hold. Then certainly (b) 1 and 2 are true.

Hence, by (b), we have $P(n)$ true for all n.

Thus (a) holds.

Hence (b) $\rightarrow$ (a).

Since we have proved (b), it follows that (a) is true. This concludes the proof of the theorem.

EXAMPLE 3.2.1 Show that:

$$1 + 2 + 3 + 4 + \ldots\ldots + n = n(n+1)/2.$$

Let $P(n)$ be the statement above that is to be proved.

Then $P(1)$ is the statement that:

$$1 = 1(2)/2.$$

This is clearly true.

If $P(n)$ holds then:

$$1 + 2 + 3 + \ldots. + n = n(n+1)/2.$$

From this we deduce that:

$$1 + 2 + 3 + 4 + \ldots. + n + (n+1) = n(n+1)/2 + (n+1).$$

The right-hand side of this equation reduces to

$$(n+1)(n+2)/2.$$

Thus we have:

$$1 + 2 + 3 + \ldots\ldots + n + (n+1) = (n+1)(n+2)/2.$$

Now this is the statement $P(n+1)$.

We conclude that $P(n) \rightarrow P(n+1)$.

By theorem 3.2.1 (a), $P(n)$ is true for all $n \in \mathbf{Z}^+$.

3.3 CONSTRUCTION OF THE RATIONALS

From **Z**, the ring of integers, we construct the rationals as follows.

Let $\mathbf{Z}^* = \mathbf{Z} - \{0\}$.

This is the set of integers leaving out the zero. We define a relation $\sim$ on the Cartesian product $\mathbf{Z} \times \mathbf{Z}^*$ by:

$$(a,b) \sim (c,d) \text{ if and only if } ad = bc;\ a, c \in \mathbf{Z};\ b, d \in \mathbf{Z}^*.$$

We leave the reader to check that $\sim$ is an equivalence relation on the set $\mathbf{Z} \times \mathbf{Z}^*$. This is just a question of verifying the 3 properties defining an equivalence relation given in section 2.14.

We denote the equivalence class containing (a,b) by $[(a,b)]$, or, more suggestively, by $[a/b]$.

Rational numbers are just these equivalence classes.

We define addition of rationals by:

$$[a/b] + [c/d] = [(ad + bc)/bd].$$

Since $b, d \in \mathbf{Z}^*$ and **Z** is an integral domain we have $bd \in \mathbf{Z}^*$ as required. However our definition may depend on which representative a/b we take in the class $[a/b]$.

For example

$$[2/3] = [4/6] = [6/9]$$

and so on, also

$$[3/4] = [6/8].$$

It is not immediately obvious that

$$[2/3] + [3/4] = [4/6] + [6/8]$$

according to our definition. This must be verified for the **general** case. We leave this to the reader.

We define multiplication of rationals by:

$$[a/b] \times [c/d] = [ac/bd].$$

All the considerations mentioned above apply. Again we leave the reader to carry out the routine checks.

The set of rationals is denoted by **Q**. Under the addition and multiplication defined above, **Q** is a **field**. Once more we leave the reader to check through the various axioms as given in section 3.1.

As usual we denote $[a/b]$ just by a/b, where it is customary to take a and b so that the highest common factor of a and b is 1.

For example

$$[2/3] = [4/6] = [6/9]$$

etc. is denoted by just 2/3.

3.4 CONSTRUCTION OF THE REALS

There are several methods of constructing the set of real numbers. We shall mention two methods in outline and give references for the interested reader to pursue.

The first method is due to Dedekind, although the essence of the idea probably goes back as far as the ancient Greeks around 400 B.C. Dedekind appears to have elaborated his theory about 1858 but did not publish it until 1872. In this theory a real number is a partition of the set of rational numbers into two subsets called the left and right classes of the partition.

For example the real number $\sqrt{2}$ is described by:

$$(L, R), \text{ where } L = \{x \in \mathbf{Q} : x^2 \leqq 2\} \text{ and } R = \{x \in \mathrm{Q} : 2 < x^2\}.$$

Addition and multiplication of these real numbers is defined next. Then the set of real numbers under the defined addition and multiplication is shown to be a field. A very clear exposition of the details of this process is given in *Calculus* by Spivak (W.A. Benjamin 1967), Chapter 28.

The second method is due to Cantor and was published in 1872 in a journal called Mathematische Annalen. In some ways this method seems more natural. A real number is regarded essentially as a sequence of rationals which 'get closer in value' to the required real number. More precisely the limit of the sequence of rationals is the required real number. A sequence like:

$$1.4,\ 1.41,\ 1.414,\ 1.4142, \ldots\ldots\ldots\ \text{defines the real number } \sqrt{2}.$$

The above is intended to give a rough idea of the process from an intuitive point of view. For a very detailed treatment see *The Real Number System in an Algebraic Setting* by J.B. Roberts (W.H. Freeman 1962). For the student who has mastered enough abstract algebra, a beautifully concise treatment is given in *Modern Algebra* by Van der Waerden (Ungar Publishing 1953). Also see *Foundations of Mathematics* by Stewart and Tall (O.U.P. 1977).

Whichever method is used, the vital properties of **R**, the set of real numbers, are summed up by saying that **R** is a **complete ordered field** under the usual operations of addition and multiplication with the usual order relation denoted by $\leqq$. The basic properties of $<$ are:

(1) For every element $a \in \mathbf{R}$ just one of the relations $a = 0$, $0 < a$, $0 < -a$ holds.

(2) If $0 < a$ and $0 < b$ then $0 < a + b$ and $0 < ab$.

To explain the idea of 'completeness' we first define upper and lower bounds.

DEFINITION 3.4.1

Let S be any non-empty subset of **R**.

Let s be **any** member of S.

Then $a \in \mathbf{R}$ is an **upper bound** of S if $s \leqq a$, and $b \in \mathbf{R}$ is a **lower bound** of S if $b \leqq s$. An upper bound of S with the extra property that all other upper bounds of S are greater is called the **least upper bound** of S. It is abbreviated to l.u.b. or sup. Similarly the greatest lower bound has the defining property that all other lower bounds of S are smaller. It is abbreviated by g.l.b or inf.

From the definition sup. and inf. are **unique, if they exist**, for a given set S. By saying that **R** is **complete** we are just asserting that sup. **does exist** for any non-empty subset S of **R** that is bounded above, and inf. exists for any non-empty subset S of **R** that is bounded below.

Note that although **Q** is an ordered field with the usual ordering it is **not** complete. To show this, consider the subset of **Q** defined by

$$S = \{x \in \mathbf{Q} : x^2 \leqq 2\}.$$

This subset is bounded above by 3/2. However S has no sup.

Note that if S is regarded as a subset of **R**, the set of real numbers, then the sup. of S is $\sqrt{2}$.

It is precisely these properties, that make **R** a complete ordered field, which enables us to do classical analysis. The reader is referred to the standard texts on analysis for further details.

3.5 THE COMPLEX NUMBERS

The complex numbers are easily defined as follows:

DEFINITION 3.5.1
The set of complex numbers is the cartesian product **R** x **R**, denoted by **C**.
Addition of complex numbers is defined by:

$$(a,b) + (c,d) = (a+c, b+d)$$

and multiplication by:

$$(a,b) \times (c,d) = (ac - bd, bc + ad).$$

As usual we leave the reader to verify that **C**, under the addition and multiplication just defined, is a field. To help in this, we remark that (0,0) is the zero, (1,0) is the multiplicative identity, $(-a,-b)$ is the additive inverse of (a,b), and $(a/(a^2+b^2), -b/(a^2+b^2))$ is the multiplicative inverse of (a,b).

Complex numbers were introduced initially to provide solutions for equations like:

$$x^2 + x + 1 = 0.$$

Essentially this reduces to giving a meaning to $\sqrt{-1}$, which is usually denoted by i, and then incorporating this into a field so that the usual arithmetic calculations can be done. One wants the smallest field which contains **R** and i. This leads to consideration of formal expressions such as $a + bi$, where $a, b \in$ **R**. But is the collection of all such expressions a well-defined set in our set theory? To avoid this difficulty and give a rigorous definition of complex numbers based on axiomatic set theory, it turns out to be most convenient to adopt Definition 3.5.1. We shall see the connection with the collection of expressions like $a + bi$ in a moment.

3.6 THE HIERARCHY OF NUMBERS

The natural numbers including the zero may be identified with the non-negative integers. This gives the inclusion or embedding **N** ⊂ **Z**. Thus **N** is regarded as a subset of **Z**.
The integer n may be identified with the rational number $[n/1]$ or (loosely) just $n/1$. This enables us to regard **Z** as a subset of **Q**. Similarly, whatever construction of the real numbers is used, we may regard the rationals as special kinds of real numbers. This allows us to take **Q** as a subset of **R**. Finally the real number a may be identified with the complex number $(a,0)$, and **R** may be considered as a subset of **C**. Thus we have the sequence of inclusions:

$$\mathbf{N} \subset \mathbf{Z} \subset \mathbf{Q} \subset \mathbf{R} \subset \mathbf{C}.$$

But do these inclusions preserve the arithmetic?

For example, if we calculate in **R** do we get the same result for a given calculation as we do when we regard the reals as special complex numbers? To help us answer this question we frame the following definition:

DEFINITION 3.6.1
Let S and T be two rings.

Let $f: S \to T$ be a function from S into T.

We call f an **homomorphism** if

$$f(a+b) = f(a) + f(b)$$

and

$$f(ab) = f(a)f(b), \text{ for all } a, b \in S.$$

If S and T also have identities we require that $f(e_S) = e_T$, where e_S is the identity in S and e_T is the identity in T.

If f is one-one in the above we call it a **monomorphism** (or sometimes an **isomorphism into**).

If f is one-one and onto (bijection) we call it an **isomorphism** and say that S and T are **isomorphic** rings.

In terms of our definitions we can pose the question raised above more precisely as follows:

(1) Let f_1: $\mathbf{Z} \to \mathbf{Q}$ be defined by $f_1(n) = [n/1], n \in \mathbf{Z}$.

Is f_1 a monomorphism?

(2) Let f_2: $\mathbf{R} \to \mathbf{C}$ be defined by $f_2(a) = (a,0), a \in \mathbf{R}$.

Is f_2 a monomorphism?

We have omitted consideration of the inclusion $\mathbf{Q} \subset \mathbf{R}$ which involves a more detailed treatment of the construction of the real numbers than we wish to consider here. We ask the reader to accept that the inclusion $\mathbf{Q} \subset \mathbf{R}$ is a monomorphism.

Now let us show that f_2 is a monomorphism.

$$f_2(a+b) = (a+b,0) = (a,0) + (b,0) = f_2(a) + f_2(b), \text{ and}$$

$$f_2(ab) = (ab,0) = (a,0) \times (b,0) = f_2(a) \times f_2(b), \text{ where}$$

we have used our definitions of addition and multiplication of complex numbers.

Now suppose that $\quad f_2(a) = f_2(b)$

Then $\quad (a,0) = (b,0).$

But the ordered pair (x,y) is equal to the ordered pair (x',y') if and only

if $\quad x = x'$ and $y = y'$.

Thus $a = b$. Hence f_2 is one-one.

Altogether f_2 is a monomorphism of **R** into **C**.

We leave the reader to verify that f_1 is a monomorphism of **Z** into **Q**.

Note that f(**R**) is a field isomorphic to **R** embedded in **C**. The operations of addition and multiplication in f(**R**) are exactly the same as those in **C**, the containing field. This situation is described by saying that f(**R**) is a **subfield** of **C**. In a similar way we say that f(**Z**) is a **subring** of **Q**.

That the inclusion $\mathbf{N} \subset \mathbf{Z}$ preserves the arithmetic of **N** we accept on an intuitive level. We have little alternative since we assumed **N** and **Z** to be intuitively known. If, however, these sets are defined within the context of set theory, then a similar analysis has to be done as for the other inclusions. Of course in this case **N** is not a ring so there are a few modifications.

3.7.1 REPRESENTATIONS

It is not very convenient to calculate with complex numbers in the form of ordered pairs of real numbers, i.e. (a,b), $a,b \in \mathbf{R}$. In this section we consider various alternative ways of representing complex numbers which facilitate their use.

(1) **Symbolic**: We consider the set S of all formal expressions of the form: $a + bi$, where a, $b \in \mathbf{R}$ and $a + bi$ is equal to $c + di$ if and only if $a = c$ and $b = d$. We define addition of these formal expressions by:

$$(a + bi) + (c + di) = (a + c) + (b + d)i,$$

and multiplication by:

$$(a + bi) \times (c + di) = (ac - bd) + (bc + ad)i.$$

In fact these expressions behave exactly like ordinary algebraic expressions involving real numbers provided we replace i^2 by -1.

Note that $bi = ib$ in the above.

It is usual to leave out the multiplication sign $\times$ and just place next to each other the two expressions which are to be multiplied.

Now define a function $f\colon \mathbf{C} \longrightarrow S$ by:

$$f(a,b) = a + bi, \text{ where } a,b \in \mathbf{R}.$$

We leave the reader to check that f is one-one and onto, and that:

$$f(\,(a,b) + (c,d)\,) = f(\,(a,b)\,) + f(\,(c,d)\,)$$

$$f(\,(a,b)\,(c,d)\,) = f(\,(a,b)\,)\; f(\,(c,d)\,).$$

Thus f is an isomorphism of $\mathbf{C}$ and S.

From this it follows that we can equally well work with the formal expressions $a + bi$ as with the ordered pairs (a,b).

EXAMPLE 3.7.1 Let us illustrate the above by calculating $(2,3)\,(3,1)^{-1}$ or, as it is usually written, $(2,3)/(3,1)$.
First using ordered pairs we have:

$$(3,1)^{-1} = (3/10, -1/10) \text{ using section 3.5.}$$

Thus $$(2,3)\,(3,1)^{-1} = (2,3)\,(3/10, -1/10) = (6/10 + 3/10, 9/10 - 2/10),$$

$$= (9/10, 7/10).$$

Using formal expressions we have:

$$\begin{aligned}(2 + 3i)/(3 + i) &= (2 + 3i)\,(3 - i)/(3 + i)\,(3 - i) = \\ &= (6 - 3i^2 + 9i - 2i)/(9 - i^2 + 3i - 3i) \\ &= (9 + 7i)/(10 + 0i) = (9 + 7i)/10 = \\ &= (9/10) + (7/10)i.\end{aligned}$$

(2) **Geometrical – the Argand Diagram:** We represent the complex number (a,b) as the point with coordinates (a,b) in the usual Euclidean plane referred to straight line axes at right angles to each other. Referred to polar coordinates r and θ we have: $a + bi = r(\cos\theta + i\sin\theta)$.

We illustrate these ideas in Figure 3.7.1

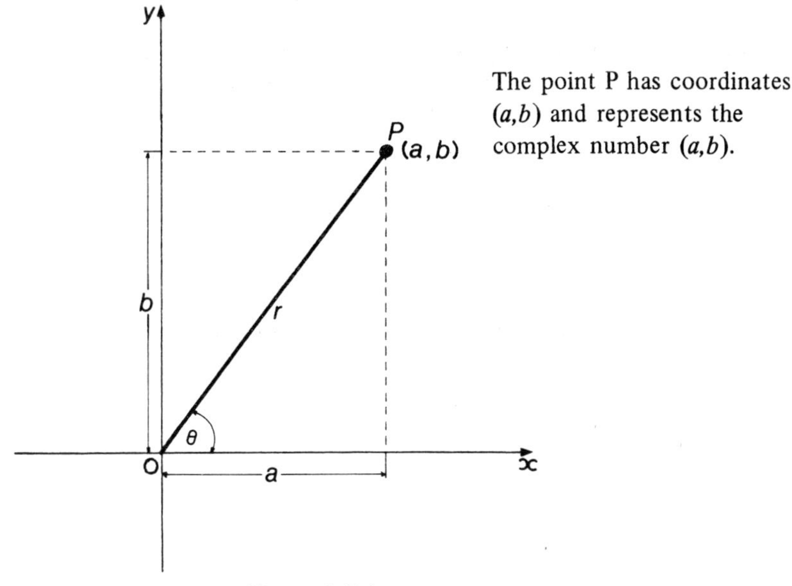

Figure 3.7.1

From the point of view of addition the complex number $(a,b) = a + bi$ may be thought of as the 2–dimensional vector $\vec{OP}$. Adding two complex numbers then corresponds exactly to the addition of the representative vectors.

3.7.2 MODULUS, ARGUMENT, AND CONJUGATE

Let $z = (a,b) = a + bi = r(\cos\theta + i\sin\theta)$

be a complex number as represented in Figure 3.7.1. We call the distance $\vec{OP}$ the **modulus** of z, and denote it by $|z|$.

Thus $\quad |z| = \sqrt{a^2 + b^2} = r.$
We call the angle θ the **argument** of z and denote it by arg z.

We usually restrict arg z to the range

$$0 \leqq \theta < 2\pi \quad \text{or} \quad -\pi < \theta \leqq \pi.$$

For obvious reasons $r(\cos\theta + i\sin\theta)$ is often called the **modulus and argument form** of the complex number.

Let $\quad z_1 = r_1(\cos\theta_1 + i\sin\theta_1)$ and $z_2 = r_2(\cos\theta_2 + i\sin\theta_2)$.

Then $\quad z_1 z_2 = r_1 r_2(\cos(\theta_1 + \theta_2) + i\sin(\theta_1 + \theta_2))$

by elementary trigonometry. Thus we deduce that:

$$|z_1 z_2| = |z_1|\,|z_2| \text{ and } \arg z_1 z_2 = \arg z_1 + \arg z_2.$$

Note that in the latter equation we may have to reduce $\arg z_1 + \arg z_2$ by a multiple of 2π in order to bring its value into the agreed range of 0 to 2π or $-\pi$ to π.

By an exactly similar argument we deduce that:

$$|z_1/z_2| = |z_1|/|z_2| \quad \text{and} \quad \arg z_1/z_2 = \arg z_1 - \arg z_2.$$

Let $z = a + bi$.

We call $a - bi$ the **conjugate** of z and denote it by $\bar{z}$. We note that:

$$z\bar{z} = |z|^2.$$

Also quite easy to check are the following useful facts:

$$\overline{z_1 + z_2} = \bar{z}_1 + \bar{z}_2, \quad \overline{z_1 z_2} = \bar{z}_1 \bar{z}_2,$$

and if

$$z = r(\cos\theta + i\sin\theta) \quad \text{then} \quad \bar{z} = r(\cos(-\theta) + i\sin(-\theta))$$

and

$$z^{-1} = 1/z = 1/r\ (\cos\theta - i\sin\theta).$$

Figure 3.7.2 illustrates the relation between z and $\bar{z}$ on the Argand diagram.

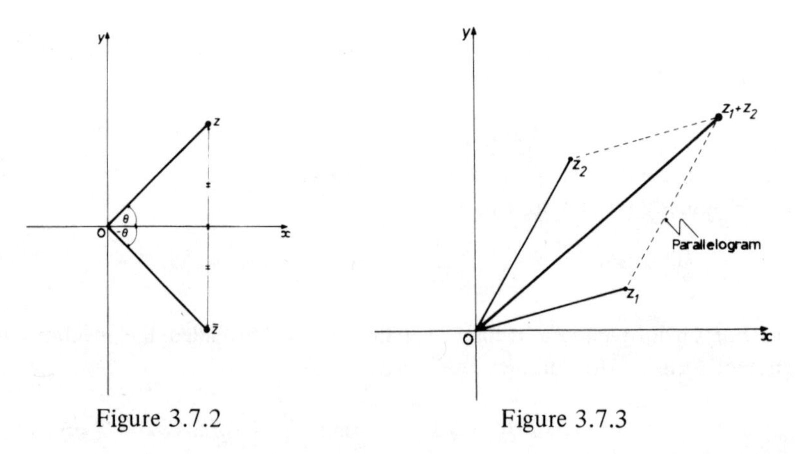

Figure 3.7.2 Figure 3.7.3

The Argand diagram may also be used to give an informal proof of the important inequality:

$$|z_1 + z_2| \leqq |z_1| + |z_2|,$$

where z_1 and z_2 are any two complex numbers.

This follows at once from Figure 3.7.3 when we recall that the sum of the lengths of two sides of a triangle is greater than the length of the third side.

Using the equation $z_1 + (z_2 - z_1) = z_2$ we deduce another useful inequality:

$$|z_2| = |z_1 + (z_2 - z_1)| \leqq |z_1| + |z_2 - z_1|,$$

by the above inequality.

Hence $\quad |z_2| - |z_1| \leqq |z_2 - z_1|.$

Interchanging z_1 and z_2 we get:

$$|z_1| - |z_2| \leqq |z_1 - z_2|.$$

However $\quad |z_1 - z_2| = |z_2 - z_1|.$

Thus $\quad \big| |z_2| - |z_1| \big| \leqq |z_2 - z_1|.$

3.8 CURVES AND REGIONS ON THE ARGAND DIAGRAM

Complex equations and inequalities may be used to specify curves and regions on the Argand diagram. Before looking at examples of this we note two useful facts.

Figure 3.8.1 illustrates the first fact.

Let z_1 and z_2 be two complex numbers.

Then $|z_2 - z_1|$ is the distance between the points representing z_1 and z_2 on the Argand diagram. This follows easily by regarding z_1 and z_2 as two-dimensional vectors in the plane.

The second fact concerns rotation of the vector z through an angle in the Argand diagram.

Let $\quad u = m(\cos\alpha + i\sin\alpha)$ and $z = r(\cos\theta + i\sin\theta)$.

Then $\quad uz = mr(\cos(\theta+\alpha) + i\sin(\theta+\alpha))$.

Thus the length of z is multiplied by $m = |u|$, and the vector z is rotated through α. In particular if $u = i$ then iz is the vector z rotated through 90 degrees.

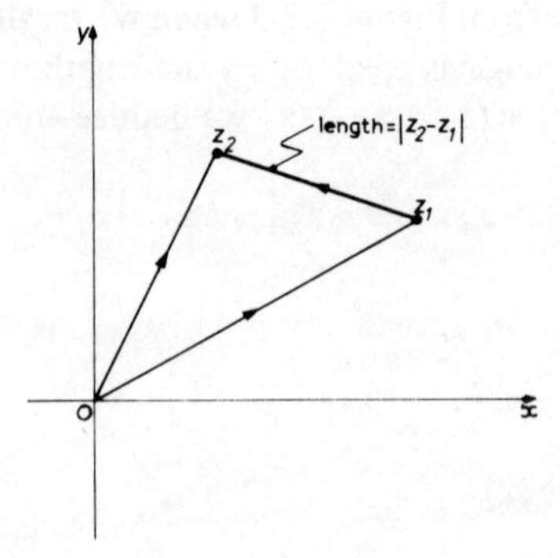

Figure 3.8.1

EXAMPLE 3.8.1 The equation $|z - z_0| = r$, where z_0 is a fixed complex number, r is a real number greater than 0, and z is a complex number which is allowed to vary subject to satisfying the above equation, tells us that the distance between z and z_0 on the Argand diagram is always r. From elementary geometry we see that as z varies it traces out a circle of radius r and centre z_0. This is shown in Figure 3.8.2

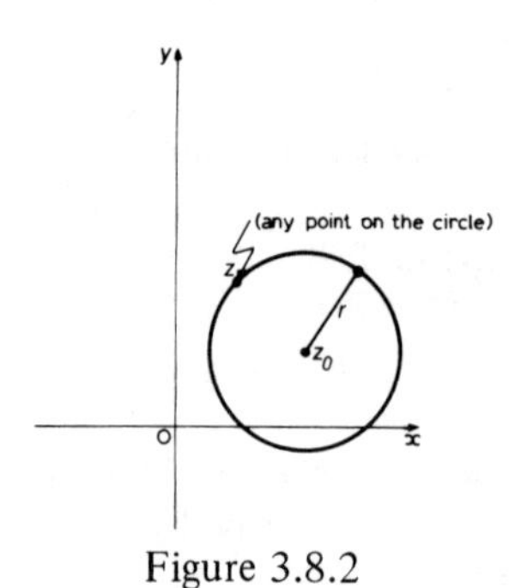

Figure 3.8.2

EXAMPLE 3.8.2 The inequality $\mathbf{R}\, z > k$, where $\mathbf{R}\, z$ denotes the real part of z, and k is real, describes the region to the right of the line $x = k$ in the Argand plane, as shown in Figure 3.8.3.

Note that if $z = a + bi$ then the real part of z is a and the imaginary part of z is b.

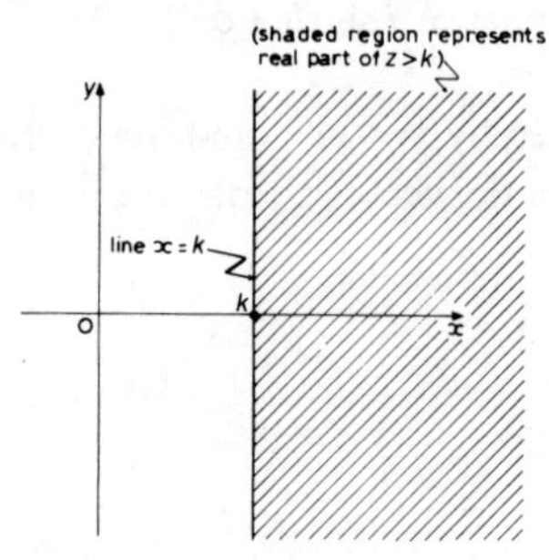

Figure 3.8.3

EXAMPLE 3.8.3 The equation $|z - 1| = |z + 1|$ tells us that the point z moves so that its distance from the point (1,0) is equal to its distance from the point (−1,0). As can be seen from Figure 3.8.4 this describes the line $x = 0$ in the Argand plane.

Calculation also gives this result as follows:

$$|(x - 1) + iy| = |(x + 1) + iy|, \text{ where } z = x + iy.$$

Thus $$(x - 1)^2 + y^2 = (x + 1)^2 + y^2,$$

$$x^2 - 2x + 1 = x^2 + 2x + 1, \text{ and finally,}$$

$$x = 0.$$

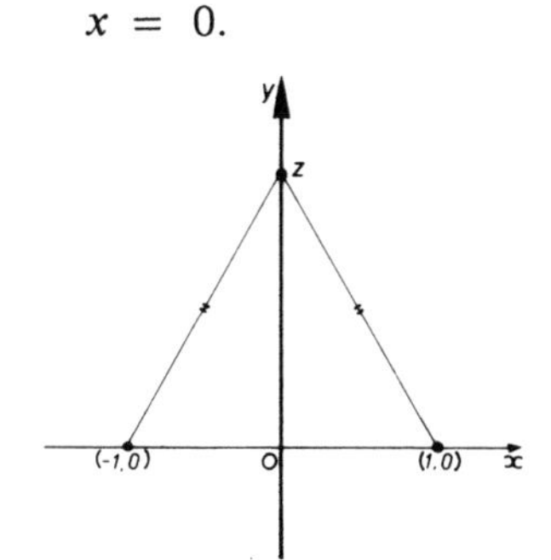

Figure 3.8.4

3.9 ROOTS OF POLYNOMIALS WITH REAL COEFFICIENTS

Let $f(z) = a_0 z^n + a_1 z^{n-1} + \ldots\ldots + a_{n-1} z + a_n$

be a polynomial with real coefficients

$$a_0, a_1, \ldots\ldots\ldots\ldots, a_n,$$

where z is a variable complex number (a complex variable).

Then $f(\bar{z}) = a_o \bar{z}^n + a_1 \bar{z}^{n-1} + \ldots\ldots + a_{n-1}\bar{z} + a_n$

$$= \overline{f(z)}, \text{ since the } a_i, i = 0,1,2,\ldots\ldots, n$$

are all real.

Thus if $f(w) = 0$

for some complex number w then

$$f(\bar{w}) = \overline{f(w)} = 0.$$

Hence $$f(\bar{w}) = 0.$$

This means that the roots of any polynomial equation with real coefficients occur in conjugate pairs.

EXAMPLE 3.9.1 Show that $1 + i$ is a root of the equation

$$z^4 - 3z^3 + 3z^2 - 2 = 0.$$

Hence solve the equation completely.

SOLUTION

$$(1+i)^4 - 3(1+i)^3 + 3(1+i)^2 - 2 =$$

$$= 1 + 4i + 6i^2 + 4i^3 + i^4 - 3(1 + 3i + 3i^2 + i^3) +$$

$$+ 3(1 + 2i + i^2) - 2 = 1 + 4i - 6 - 4i + 1 - 3 - 9i + 9 +$$

$$+ 3i + 3 + 6i - 3 - 2 = 0.$$

Thus $1 + i$ is a root of the equation.

By the above result, $1 - i$ is also a root.

Thus $(z - (1 + i))\ (z - (1 - i))$

is a factor of

$$z^4 - 3z^3 + 3z^2 - 2.$$

Factorizing we get:

$$z^4 - 3z^3 + 3z^2 - 2 = (z^2 - 2z + 2)\ (z^2 - z - 1) = 0.$$

The remaining roots of the equation are the roots of:

$$z^2 - z - 1 = 0.$$

By the usual methods for solving a quadratic equation, these roots are:

$$(1 + \sqrt{5})/2 \qquad \text{and} \qquad (1 - \sqrt{5})/2.$$

Hence the 4 roots of the equation

$$z^4 - 3z^3 + 3z^2 - 2 = 0 \text{ are:}$$

$$1 + i,\ 1 - i,\ (1 + \sqrt{5})/2,\ (1 - \sqrt{5})/2.$$

3.10 EXTRACTION OF SQUARE ROOTS

Let $z = a + ib$.

We want to find w so that $w^2 = z$.

Let $w = x + iy$.

Then $(x + iy)^2 = a + ib$.

Thus $(x^2 - y^2) + 2ixy = a + ib$.

Hence $x^2 - y^2 = a$ and $2xy = b$.

Thus $x^2 - b^2/4x^2 = a$.

Solving for x^2, we get:

$$x^2 = (a \pm \sqrt{a^2 + b^2})/2.$$

Since $a - \sqrt{a^2 + b^2}$

is negative while x^2 is positive, we have just

$$x^2 = (a + \sqrt{a^2 + b^2})/2.$$

Hence $x = \pm\sqrt{(a + \sqrt{a^2 + b^2})/2}$.

Then $y = b/2x = b/\pm 2\sqrt{(a + \sqrt{a^2 + b^2})/2}$.

For general nth roots we use an interesting result called De Moivre's theorem which we consider in the next section.

3.11 DE MOIVRE'S THEOREM

THEOREM 3.11.1

Let p and $q \neq 0$ be integers.

Then p/q is an arbitrary rational number and **one** value of

$$(\cos\theta + i\sin\theta)^{p/q} \text{ is } \cos\tfrac{p}{q}\theta + i\sin\tfrac{p}{q}\theta.$$

PROOF The proof is in three main parts. First for positive integers, next for negative integers, finally for rational numbers.

Let n be a positive integer.

Let $P(n)$ be the statement of the theorem.

Then $P(1)$ states that:

$$(\cos\theta + i\sin\theta)^1 = \cos\theta + i\sin\theta.$$

Thus $P(1)$ is true.

Now suppose $P(n)$ is true. Then

$$\begin{aligned}(\cos\theta + i\sin\theta)^{n+1} &= (\cos\theta + i\sin\theta)^n(\cos\theta + i\sin\theta)\\ &= (\cos n\theta + i\sin n\theta)(\cos\theta + i\sin\theta)\\ &= (\cos n\theta\cos\theta - \sin n\theta\sin\theta) + \\ &\quad + i(\cos n\theta\sin\theta + \cos\theta\sin n\theta)\\ &= \cos(n+1)\theta + i\sin(n+1)\theta,\end{aligned}$$

by elementary trigonometry.

Thus $P(n+1)$ is true.

Hence $P(n) \rightarrow P(n+1)$.

The theorem follows by induction for all positive integers n.

For $n = 0$ we have

$$(\cos\theta + i\sin\theta)^0 = 1 = \cos 0 + i\sin 0$$

which is true.

For n a negative integer put

$$n = -m \text{ with } m > 0.$$

Then
$$\begin{aligned}(\cos\theta + i\sin\theta)^n &= ((\cos\theta + i\sin\theta)^m)^{-1}\\ &= (\cos m\theta + i\sin m\theta)^{-1}\end{aligned}$$

$$= \cos m\,\theta - i \sin m\,\theta$$

$$= \cos(-m)\theta + i \sin(-m)\theta$$

$$= \cos n\,\theta + i \sin n\,\theta.$$

This proves the theorem for negative integers.

Finally let p and $q \neq 0$ be integers. We have:

$$(\cos {}^{p}\!/\!_{q}\,\theta + i \sin {}^{p}\!/\!_{q}\,\theta)^q = \cos p\,\theta + i \sin p\,\theta$$

$$= (\cos\theta + i \sin\theta)^p.$$

Hence $(\cos {}^{p}\!/\!_{q}\,\theta + i \sin {}^{p}\!/\!_{q}\,\theta) = (\cos\theta + i \sin\theta)^{p/q}.$

This completes the proof of the theorem.

Note that $(\cos\theta + i \sin\theta)^{p/q}$ has q possible values, namely the q qth roots of $(\cos\theta + i\sin\theta)^p$; see section 3.13.

By regarding a real number as a limit of a sequence of rationals it is possible to extend the above theorem to real exponents. However by assuming some results from analysis we can argue more easily as follows. We have:

$$e^z = \exp z = 1 + z/1! + z^2/2! + z^3/3! + \ldots\ldots + z^n/n! + \ldots.$$

Put $z = i\,\theta.$

Then $e^{i\theta} = (1 - \theta^2/2! + ..) + i(\theta - \theta^3/3! + \ldots.)$

assuming result on power series etc.

Now $\cos\theta = 1 - \theta^2/2! + \ldots\ldots$ and

$\sin\theta = \theta - \theta^3/3! + \ldots\ldots$

Hence $e^{i\theta} = \exp i\,\theta = \cos\theta + i \sin\theta.$

Then $(\cos\theta + i\sin\theta)^n = (e^{i\theta})^n = e^{in\theta} = \cos n\,\theta + i \sin n\,\theta.$

This applies for any real n.

We can write:

$$z = r(\cos\theta + i\sin\theta) = re^{i\theta},$$

$$\cos\theta = (e^{i\theta} + e^{-i\theta})/2 \quad \text{and} \quad \sin\theta = (e^{i\theta} - e^{-i\theta})/2i.$$

3.12 ROOTS OF UNITY

The solutions to the equation $z^n = 1$ turn out to be very important in advanced mathematics. Here we are assuming that n is a positive integer.
With this restriction the solutions to $z^n = 1$ are called the nth roots of unity.

We write $$z^n = \cos 2m\pi + i \sin 2m\pi.$$

Then $$z = (\cos 2m\pi + i \sin 2m\pi)^{1/n} = \cos 2m\pi/n + i \sin 2m\pi/n$$

by De Moivre's theorem.

Now take $$m = 0,1,2,3,4, \ldots\ldots, (n-1)$$

in succession to obtain n distinct roots.

On the Argand diagram these roots are spaced round a circle of unit radius (usually called the unit circle) at angular intervals of $2\pi/n$ as shown in Figure 3.12.1

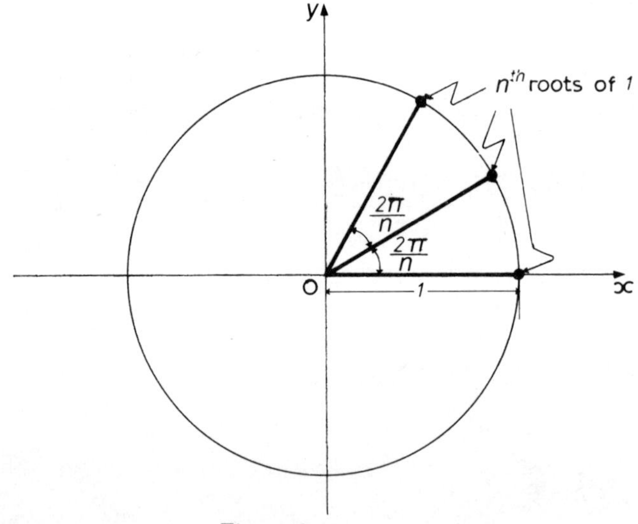

Figure 3.12.1

$$u = \cos 2\pi/n + i \sin 2\pi/n$$

is called a **primitive** nth root of unity, because $u^t \neq 1$ if $t < n$ and so all the other nth roots of unity are powers of u. In fact the n nth roots of unity are:

$$u, u^2, u^3, u^4, \ldots\ldots\ldots, u^{n-1}, u^n = 1.$$

3.13 ROOTS OF AN ARBITRARY COMPLEX NUMBER a.

A similar procedure gives the n nth roots of any complex number a. These are the solutions of the equation $z^n = a$, where $a = r(\cos\theta + i \sin\theta)$ in modulus and argument form.

Writing a in general form:

$$a = r(\cos(\theta + 2m\pi) + i\sin(\theta + 2m\pi))$$

we have $z = a^{1/n} = r^{1/n}(\cos(\theta + 2m\pi)/n + i\sin(\theta + 2m\pi)/n)$,

using De Moivre's theorem.

Putting $m = 0,1,2,3,\ldots\ldots,(n-1)$

in turn in the above, we get the required n nth roots of a. In the above $r^{1/n}$ is the arithmetical nth root of r.

Thus $\quad r^{1/n} = e^{1/n \log r}$.

As for the nth roots of unity, the n nth roots of a are spaced round a circle in the Argand diagram.
This time the radius of the circle is $r^{1/n}$ but the angular interval is the same $2\pi/n$. The first root is displaced from the real axis by an angle of θ/n.

This is shown in Figure 3.13.1

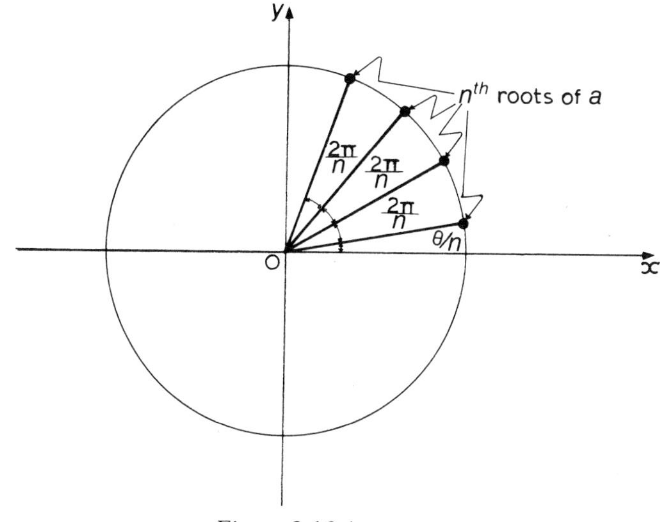

Figure 3.13.1

3.14 FACTORISATION OF POLYNOMIALS INTO REAL FACTORS

Rather than become involved in the complexities of a general treatment, we shall introduce the ideas by considering a special case which can easily be generalised.

Let us suppose that we want to factorise $z^6 + z^3 + 1$ into polynomial factors with real coefficients.

Put $\quad u = z^3$.

Then $z^6 + z^3 + 1 = u^2 + u + 1$.

We now solve the equation

$$u^2 + u + 1 = 0.$$

This has roots

$$(-1 \pm \sqrt{3}\, i)/2.$$

Thus $z^3 = (-1 + \sqrt{3}\, i)/2$ or $z^3 = (-1 - \sqrt{3}\, i)/2$.

We solve these equations by the methods of section 3.13. We have:

$$z^3 = \cos(2\pi/3 + 2m\pi) \pm i \sin(2\pi/3 + 2m\pi).$$

$$\text{Thus } z = \cos(6m + 2)\,\pi/9 \pm i \sin(6m + 2)\,\pi/9,$$

where m takes on the values 0,1,2 in turn. The roots of $z^6 + z^3 + 1 = 0$ are therefore given by:

$$z_1 = \cos 2\pi/9 + i \sin 2\pi/9,\ \overline{z}_1 = \cos 2\pi/9 - i \sin 2\pi/9,$$

$$z_2 = \cos 8\pi/9 + i \sin 8\pi/9,\ \overline{z}_2 = \cos 8\pi/9 - i \sin 8\pi/9,$$

$$z_3 = \cos 14\pi/9 + i \sin 14\pi/9,\ \overline{z}_3 = \cos 14\pi/9 - i \sin 14\pi/9.$$

Hence $z^6 + z^3 + 1$ factorises as:

$$(z - z_1)(z - \overline{z}_1)(z - z_2)(z - \overline{z}_2)(z - z_3)(z - \overline{z}_3) \equiv$$

$$(z^2 - (z_1 + \overline{z}_1)z + |z_1|^2)(z^2 - (z_2 + \overline{z}_2)z + |z_2|^2)(z - (z_3 + \overline{z}_3)z + |z_3|^2$$

where conjugate factors have been coupled together.

Putting in the actual values we have:

$$z^6 + z^3 + 1 \equiv$$

$$(z^2 - (2\cos 2\pi/9)z + 1)\ (z^2 - (2\cos 8\pi/9)z + 1)\ (z^2 - (2\cos 14\pi/9)z + 1$$

We can also get trigonometric identities from this type of factorisation. For example: divide the above identity by z to get

$$z^3 + 1 + 1/z^3 \equiv$$

$(z - 2\cos 2\pi/9 + 1/z)\ (z - 2\cos 8\pi/9 + 1/z)\ (z - 2\cos 14\pi/9 + 1/z)$.

Now put $z = e^{i\theta}$.

Then $z^3 + 1/z^3 = 2\cos 3\theta$, and $z + 1/z = 2\cos\theta$.

Thus $2\cos 3\theta + 1 \equiv 8(\cos\theta - \cos 2\pi/9)(\cos\theta - \cos 8\pi/9)(\cos\theta - \cos 14\pi/9$

EXERCISES

1. We define a relation $\equiv$ on the set of integers $\mathbf{Z}$ as follows: $a \equiv b$ if and only if 6 divides $(a - b)$, which is usually written $6 \mid (a - b)$, where $a, b \in \mathbf{Z}$.

Prove

(a) $\equiv$ is an equivalence relation on $\mathbf{Z}$;

(b) if $\bar{a}$ denotes the equivalence class containing the integer a, then $\bar{a} + \bar{b} = \overline{(a + b)}$ and $\bar{a} \cdot \bar{b} = \overline{(ab)}$ define valid operations of addition and multiplication on the set $\mathbf{Z}_6$ of equivalence classes;

(c) $\mathbf{Z}_6$, with addition and multiplication defined as above, is a commutative ring with an identity;

(d) $\mathbf{Z}_6$ contains two non-zero elements whose product is zero i.e. zero-divisors;

(e) $\mathbf{Z}_6$ is not a field.

2. Repeat question 1 above with 5 in place of 6, and in place of parts (d) and (e), prove that (d) $\mathbf{Z}_5$ is a field. (See also Theorem 7.4.2).

3. Prove by induction that

$$1^2 + 2^2 + 3^2 + \dots\dots\dots + n^2 = n(n + 1)(2n + 1)/6$$

for all positive integers n.

4. Prove by induction on n that

$$1^3 + 2^3 + \dots\dots\dots + n^3 = (1 + 2 + \dots\dots + n)^2$$

for all positive integers n.

5. Solve for z in the equations:

(a) $(2+i)z+i=3$,

(b) $(z-1)/(z-i)=2/3$,

(c) $z^2-(3+i)z+4+3i=0$,

(d) $z^4-2z^2+4=0$.

6. Find the modulus of

$$(1+2i)^{12}/(1-2i)^9.$$

7. Verify that the points

$$2+i,\ 3+2i,\ 2+3i,\ 1+2i$$

are the vertices of a square on the Argand diagram.

8. Find the region in the z-plane described by the inequality

$$|z-1|+|z-i|\leqq 4.$$

9. If

$$(z-2i)/(2z-1)$$

is purely imaginary, prove that the locus of z in the Argand diagram is a circle. Find the centre and radius.

10. If

$$(z-6i)/(z+8)$$

is real, prove that the locus of z in the Argand diagram is a straight line.

11. Show that if

$\lambda \neq 0, 1$, then the equation $|(z-a)/(z-b)| = \lambda$

represents a circle which contains a or b in its interior according as

$\lambda < 1$ or $\lambda > 1$.

12. Prove that

$$(1 + \sin\theta + i\cos\theta)/(1 + \sin\theta - i\cos\theta) = \sin\theta + i\cos\theta.$$

Hence show that

$$(1 + \sin {}^{\pi}/5 + i\cos {}^{\pi}/5)^5 + i\,(1 + \sin {}^{\pi}/5 - i\cos {}^{\pi}/5)^5$$

13. Find all the sixth roots of unity in the form $x + iy$ with x and y both real.

14. Calculate $(2 + 2i)^{1/3}$.

15. Show that if $|z| = 1$ and the real part of z is $-1/2$, then $z^3 = 1$.

16. Show that the complex numbers 0, u, v correspond to the vertices of an equilateral triangle in the Argand diagram if and only if $u^2 + v^2 = uv$.

Hence show that the numbers z_1, z_2, z_3 correspond to the vertices of an equilateral triangle if and only if

$$z_1^2 + z_2^2 + z_3^2 = z_2 z_3 + z_3 z_1 + z_1 z_2.$$

17. Find the roots of $z^5 + 1 = 0$ and indicate them on an Argand diagram.

18. Resolve $z^5 + 1$ into linear and quadratic factors with real coefficients. Deduce that

$$4 \sin {}^{\pi}/10 \cos {}^{\pi}/5 = 1.$$

19. Show that the roots of

$$(z - 1)^6 + (z + 1)^6 = 0$$

are $\pm i \cot {}^{\pi}/12, \quad \pm i \cot {}^{5\pi}/12, \quad \pm i.$

20. By use of De Moivre's theorem, show that

$$\cos 4\theta = 8\cos^4\theta - 8\cos^2\theta + 1.$$

21. Express $\cos^6\theta$ in terms of cosines of integral multiples of θ.

CHAPTER 4

Matrices, Determinants, Vectors and Linear Transformations

4.1 MATRICES

Systems of simultaneous linear equations such as:

$$a_{11}x_1 + a_{12}x_2 + a_{13}x_3 + a_{14}x_4 = k_1$$

$$a_{21}x_1 + a_{22}x_2 + a_{23}x_3 + a_{24}x_4 = k_2$$

$$a_{31}x_1 + a_{32}x_2 + a_{33}x_3 + a_{34}x_4 = k_3$$

or a linear change of coordinates in geometry as follows:

$$x_1' = a_{11}x_1 + a_{12}x_2 + a_{13}x_3$$

$$x_2' = a_{21}x_1 + a_{22}x_2 + a_{23}x_3$$

$$x_3' = a_{31}x_1 + a_{32}x_2 + a_{33}x_3$$

both involve an array of coefficients of the form:

$$\begin{matrix} a_{11} & a_{12} & a_{13} & a_{14} \\ a_{21} & a_{22} & a_{23} & a_{24} \\ a_{31} & a_{32} & a_{33} & a_{34} \end{matrix} \quad \text{or} \quad \begin{matrix} a_{11} & a_{12} & a_{13} \\ a_{21} & a_{22} & a_{23} \\ a_{31} & a_{32} & a_{33} \end{matrix}$$

In working with these systems, the English mathematician Cayley and others in the 1850's found it convenient to use a shorthand notation. (In fact Cayley wrote a paper on this in 1858).

They wrote the above system of equations in the form:

$$Ax = k, \text{ where } A = \begin{pmatrix} a_{11} & a_{12} & a_{13} & a_{14} \\ a_{21} & a_{22} & a_{23} & a_{24} \\ a_{31} & a_{32} & a_{33} & a_{34} \end{pmatrix}$$

$$x = \begin{pmatrix} x_1 \\ x_2 \\ x_3 \\ x_4 \end{pmatrix}, \quad \text{and} \quad k = \begin{pmatrix} k_1 \\ k_2 \\ k_3 \end{pmatrix}.$$

The transformation became: $\quad x' = Bx$, where

$$B = \begin{pmatrix} a_{11} & a_{12} & a_{13} \\ a_{21} & a_{22} & a_{23} \\ a_{31} & a_{32} & a_{33} \end{pmatrix}, \quad x' = \begin{pmatrix} x_1' \\ x_2' \\ x_3' \end{pmatrix}, \text{ and } \quad x = \begin{pmatrix} x_1 \\ x_2 \\ x_3 \end{pmatrix}.$$

We call A a **matrix** and $x = \begin{pmatrix} x_1 \\ x_2 \\ x_3 \\ x_4 \end{pmatrix}$ a **vector**;

the latter may be regarded as a special kind of matrix, see later.

$$\text{The matrix } A = \begin{pmatrix} a_{11} & a_{12} & \cdots\cdots\cdots & a_{1n} \\ a_{21} & a_{22} & \cdots\cdots\cdots & a_{2n} \\ \cdot & & & \\ \cdot & & & \\ \cdot & & & \\ a_{m1} & a_{m2} & \cdots\cdots\cdots & a_{mn} \end{pmatrix}, \text{ where}$$

an horizontal line of elements is called a **row** and a vertical line of elements is called a **column**, is often denoted by (a_{ij}), where a_{ij} is the element in the ith row and the jth column.
We also use the symbol $(A)_{ij}$ to denote a_{ij}.

In order to use the shorthand notation in calculations, an arithmetic of matrices and vectors was developed with the following operations.

Addition. $\quad A + B = (a_{ij} + b_{ij})$, that is $(A + B)_{ij} = a_{ij} + b_{ij}$, where $\quad A = (a_{ij})$ and $B = (b_{ij})$.

Here a_{ij} refers to the element in the ith row (horizontal) and jth column (vertical) of the matrix. These elements will usually be real numbers but in the general theory may well be elements from some abstract ring. If no specific mention is made of the nature of these elements, the reader may assume that they are real numbers.

Note that for + to be defined A and B must have the same number of rows and the same number of columns, e.g.

$$\begin{pmatrix} 1 & 2 & 1 \\ 0 & -1 & 2 \end{pmatrix} + \begin{pmatrix} 0 & -1 & 2 \\ 4 & 1 & 3 \end{pmatrix} = \begin{pmatrix} 1+0 & 2-1 & 1+2 \\ 0+4 & -1+1 & 2+3 \end{pmatrix}, \text{ but}$$

$$\begin{pmatrix} 5 & 0 \\ 2 & -1 \end{pmatrix} + \begin{pmatrix} 3 & 6 & 1 \\ 2 & 0 & 3 \end{pmatrix} \quad \text{is **not** defined.}$$

Multiplication by a scalar. The name **scalar** is given to the elements of a matrix. If a is a scalar and A is a matrix we define

$$aA \text{ to be } (aa_{ij}); \text{ that is } (aA)_{ij} = aa_{ij}.$$

Product. We want BA, the product of B and A in that order, to be the matrix with the property that if

$$\mathbf{x}'' = B\mathbf{x}' \quad \text{and} \quad \mathbf{x}' = A\mathbf{x}$$

then $\mathbf{x}'' = (BA)\mathbf{x}$.

This leads to the definition:

$$BA = (c_{rs}), \text{ where } c_{rs} = \sum_{i=1}^{n} b_{ri}\, a_{is},$$

and number of columns of B equals number of rows of A. For example:

$$\begin{pmatrix} 2 & 1 & 3 \\ 0 & -1 & 2 \end{pmatrix} \begin{pmatrix} 1 & 2 \\ 0 & 1 \\ -1 & 3 \end{pmatrix} = \begin{pmatrix} 2\times 1 + 1\times 0 + 3\times(-1) & 2\times 2 + 1\times 1 + 3\times 3 \\ 0\times 1 + (-1)\times 0 + 2\times(-1) & 0\times 2 + (-1)\times 1 + 2\times 3 \end{pmatrix}$$

$$= \begin{pmatrix} -1 & 14 \\ -2 & 5 \end{pmatrix}, \text{ but}$$

$$\begin{pmatrix} 2 & 1 & 3 \\ 0 & -1 & 2 \end{pmatrix} \begin{pmatrix} 1 & 2 \\ 0 & 1 \end{pmatrix} \quad \text{is **not** defined.}$$

Further illustrations are given in Examples 4.1.1.

The above definitions apply to vectors when they are regarded as matrices with either n rows and one column or n columns and one row.

In general a matrix with m rows and n columns is called an $m \times n$ matrix or a matrix of **order** $m \times n$. A matrix of order $n \times n$ is called a square matrix of degree n.

Transpose.

Let $A = (a_{ij})$. Let $B = (b_{ij})$.

If $b_{rs} = a_{sr}$, we call B the transpose of A and write $B = A'$.

Of course A is similarly the transpose of B.

Thus $A = B'$.

We have the following results.

THEOREM 4.1.1

(1) $A'' = A$.

(2) $(AB)' = B'A'$.

PROOF

(1) If $B = A'$ then $A = B'$, thus $A = B' = A''$.

(2) Let $A = (a_{ij})$ and $B = (b_{ij})$. Then $AB = (c_{rs})$

where $c_{rs} = \sum_{i=1}^{n} a_{ri} b_{is}$.

Let $(AB)' = (d_{sr})$ where $d_{sr} = c_{rs}$.

From the definition we have

$B' = (b'_{ij})$ and $A' = (a'_{ij})$ with $b'_{ij} = b_{ji}$ and $a'_{ij} = a_{ji}$.

Thus $B'A' = (d'_{sr})$, where

$$d'_{sr} = \sum_{i=1}^{n} b'_{si} a'_{ir} = \sum_{i=1}^{n} b_{is} a_{ri} = \sum_{i=1}^{n} a_{ri} b_{is} = c_{rs} = d_{sr}.$$

Thus $d'_{sr} = d_{sr}$.

Hence $B'A' = (d_{sr}) = (AB)'$.

In **section** 3.1 we gave the definitions of various kinds of ring. **R** the field of real numbers, or **C** the field of complex numbers provide us with examples of infinite fields. The fields $\mathbf{Z}_p$, for p prime, are examples of finite fields. $\mathbf{Z}_n$, n not prime, is an example of a commutative ring with zero divisors; see exercises 3.1 and 3.2 for details. **Z** is an integral domain and the subring of even integers is a commutative ring without an identity. However, as yet, we have no example of a non-commutative ring.

The following theorem fills the gap.

THEOREM 4.1.2

The set $M_n(F)$ of all $n \times n$ matrices, whose elements come from any field F, with addition and multiplication defined as above, is a non-commutative ring with an identity.

PROOF

The zero of $M_n(F)$ is the so-called **null** or **zero** matrix in which all entries are zero. The additive inverse of

$$A = (a_{ij}) \quad \text{is} \quad -A = (-a_{ij}).$$

The multiplicative identity is the so-called identity matrix I_n which has the identity of F at each position of the leading diagonal and the zero of F elsewhere. For example:

$$I_3 = \begin{pmatrix} 1 & 0 & 0 \\ 0 & 1 & 0 \\ 0 & 0 & 1 \end{pmatrix} .$$

The only axioms which present any difficulty in checking are the associativity of the multiplication and the distributive rules.

Let $A = (a_{ij})$, $B = (b_{ij})$ and $C = (c_{ij})$.

Then the entry in the ith row and jth column of $A(BC)$ denoted by $[A(BC)]_{ij}$ is:

$$\sum_{r=1}^{n} a_{ir}(BC)_{rj} = \sum_{r=1}^{n} a_{ir}\left(\sum_{k=1}^{n} b_{rk}\, c_{kj}\right) = \sum_{r=1}^{n} \sum_{k=1}^{n} a_{ir}(b_{rk}\, c_{kj}).$$

Similarly $[(AB)C]_{ij}$ is:

$$\sum_{k=1}^{n} (AB)_{ik}\, c_{kj} = \sum_{k=1}^{n} (\sum_{r=1}^{n} a_{ir}\, b_{rk})\, c_{kj} = \sum_{k=1}^{n} \sum_{r=1}^{n} (a_{ir}\, b_{rk})\, c_{kj}.$$

By associativity in F we have the required result that

$$[A(BC)]_{ij} = [(AB)C]_{ij} \text{ and hence } A(BC) = (AB)C.$$

For the distributive rules we argue as follows:

$$[A(B+C)]_{ij} = \sum_{r=1}^{n} a_{ir}(b_{rj} + c_{rj}) = \sum_{r=1}^{n} a_{ir}\, b_{rj} + \sum_{r=1}^{n} a_{ir}\, c_{rj}$$
$$= (AB)_{ij} + (AC)_{ij}.$$

Hence:

$$A(B+C) = AB + AC.$$

Similarly we have:

$$(B+C)A = BA + CA.$$

That the ring $M_n(F)$ is non-commutative may be seen by taking

$$A = \begin{pmatrix} 0 & 0 & 0 \\ 1 & 0 & 0 \\ 1 & 1 & 0 \end{pmatrix} \quad \text{and} \quad B = \begin{pmatrix} 0 & 1 & 1 \\ 0 & 0 & 1 \\ 0 & 0 & 0 \end{pmatrix}$$

in the case when n is 3. Similarly constructed $n \times n$ matrices A and B demonstrate the same result in the general case.

Special Matrices. We have noted above the identity matrix I_n and the null or zero matrix O_n. Other special matrices are as follows.

$$aI_n = \begin{pmatrix} a & 0 & 0 \ldots\ldots\ldots 0 \\ 0 & a & 0 \ldots\ldots\ldots 0 \\ 0 & 0 & \quad\quad\quad . \\ . & & \quad\quad\quad . \\ . & & \quad\quad\quad . \\ 0 \ldots & \ldots & \ldots\ldots\ldots a \end{pmatrix}$$ is called a **scalar** matrix.

A is said to be **symmetric** if $A = A'$ and **skew-symmetric** if $A = -A'$. If

$$A \in M_n(\mathbf{C}) \text{ then } \bar{A} = (\bar{a}_{ij}).$$

We then say that A is **Hermitian** if $A = \bar{A}'$ and **skew-Hermitian** if $A = -\bar{A}'$. Hermitian matrices are of importance in quantum mechanics.

EXAMPLES 4.1.1

(1) $$\begin{pmatrix} 1 & 2 & 1 \\ 0 & 1 & 2 \\ -2 & 0 & 1 \end{pmatrix} \begin{pmatrix} 0 & 1 \\ 1 & -1 \\ 0 & 1 \end{pmatrix} = \begin{pmatrix} 2 & 0 \\ 1 & 1 \\ 0 & -1 \end{pmatrix}.$$

But notice that:

$$\begin{pmatrix} 0 & 1 \\ 1 & -1 \\ 0 & 1 \end{pmatrix} \begin{pmatrix} 1 & 2 & 1 \\ 0 & 1 & 2 \\ -2 & 0 & 1 \end{pmatrix}$$

is not defined.

(2) $$\begin{pmatrix} 1 & 0 & -5 \\ 0 & -2 & 4 \\ -5 & 4 & 1 \end{pmatrix}$$ is symmetric.

$$\begin{pmatrix} 0 & 2 & 1 \\ -2 & 0 & -6 \\ -1 & 6 & 0 \end{pmatrix}$$ is skew-symmetric.

$$\begin{pmatrix} -2 & i & 0 \\ -i & 0 & 1+i \\ 0 & 1-i & 4 \end{pmatrix}$$ is Hermitian.

$$\begin{pmatrix} 0 & 1+i & 2 \\ -1+i & 0 & 2-i \\ -2 & -2-i & 0 \end{pmatrix}$$ is skew-Hermitian.

(3) $M_2(\mathbf{Z}_2)$ is a non-commutative ring with identity with 16 members. In detail these are:

$$\begin{pmatrix} 0 & 0 \\ 0 & 0 \end{pmatrix}, \begin{pmatrix} 0 & 1 \\ 0 & 0 \end{pmatrix}, \begin{pmatrix} 0 & 0 \\ 0 & 1 \end{pmatrix}, \begin{pmatrix} 0 & 0 \\ 1 & 0 \end{pmatrix},$$

$$\begin{pmatrix} 1 & 0 \\ 0 & 0 \end{pmatrix}, \begin{pmatrix} 1 & 1 \\ 0 & 0 \end{pmatrix}, \begin{pmatrix} 1 & 0 \\ 0 & 1 \end{pmatrix}, \begin{pmatrix} 1 & 0 \\ 1 & 0 \end{pmatrix},$$

$$\begin{pmatrix} 0 & 1 \\ 0 & 1 \end{pmatrix}, \begin{pmatrix} 0 & 1 \\ 1 & 0 \end{pmatrix}, \begin{pmatrix} 0 & 0 \\ 1 & 1 \end{pmatrix}, \begin{pmatrix} 0 & 1 \\ 1 & 1 \end{pmatrix},$$

$$\begin{pmatrix} 1 & 0 \\ 1 & 1 \end{pmatrix}, \begin{pmatrix} 1 & 1 \\ 1 & 0 \end{pmatrix}, \begin{pmatrix} 1 & 1 \\ 0 & 1 \end{pmatrix}, \begin{pmatrix} 1 & 1 \\ 1 & 1 \end{pmatrix}.$$

(4) $\begin{pmatrix} -3 & 0 \\ 0 & -3 \end{pmatrix}$ is a scalar matrix.

(5) Let $A = \begin{pmatrix} 1 & -1 \\ 0 & 1 \\ 3 & 0 \end{pmatrix}$, $B = \begin{pmatrix} 2 & 0 \\ 0 & -1 \end{pmatrix}$, $C = \begin{pmatrix} -1 & 2 \\ 3 & -2 \end{pmatrix}$.

Then we have:

$$A(B+C) = \begin{pmatrix} 1 & -1 \\ 0 & 1 \\ 3 & 0 \end{pmatrix} \begin{pmatrix} 1 & 2 \\ 3 & -3 \end{pmatrix} = \begin{pmatrix} -2 & 5 \\ 3 & -3 \\ 3 & 6 \end{pmatrix}$$

$$AB + AC = \begin{pmatrix} 1 & -1 \\ 0 & 1 \\ 3 & 0 \end{pmatrix} \begin{pmatrix} 2 & 0 \\ 0 & -1 \end{pmatrix} + \begin{pmatrix} 1 & -1 \\ 0 & 1 \\ 3 & 0 \end{pmatrix} \begin{pmatrix} -1 & 2 \\ 3 & -2 \end{pmatrix}$$

$$= \begin{pmatrix} 2 & 1 \\ 0 & -1 \\ 6 & 0 \end{pmatrix} + \begin{pmatrix} -4 & 4 \\ 3 & -2 \\ -3 & 6 \end{pmatrix} = \begin{pmatrix} -2 & 5 \\ 3 & -3 \\ 3 & 6 \end{pmatrix}$$

Note that $A + B$ is not defined.

4.2 DETERMINANTS

Consider the solution of the simultaneous equations

$$Ax = k \text{ where } A = \begin{pmatrix} a_{11} & a_{12} \\ a_{21} & a_{22} \end{pmatrix}, \; x = \begin{pmatrix} x_1 \\ x_2 \end{pmatrix}, \text{ and } k = \begin{pmatrix} k_1 \\ k_2 \end{pmatrix}.$$

The solution is:

$$x_1 = (k_1a_{22} - k_2a_{12})/(a_{11}a_{22} - a_{12}a_{21}),$$

$$x_2 = (k_2a_{11} - k_1a_{21})/(a_{11}a_{22} - a_{12}a_{21}).$$

It is convenient to have a symbol for the denominator of these expressions, which depends only on the matrix A. This dependence is reflected by calling it the determinant of the matrix A and denoting it by the symbol $|A|$.

Thus $$|A| = \begin{vmatrix} a_{11} & a_{12} \\ a_{21} & a_{22} \end{vmatrix} = (a_{11}a_{22} - a_{12}a_{21}).$$

Similarly in solving the simultaneous equations

$$Ax = k \text{ where } A = \begin{pmatrix} a_{11} & a_{12} & a_{13} \\ a_{21} & a_{22} & a_{23} \\ a_{31} & a_{32} & a_{33} \end{pmatrix}, \quad x = \begin{pmatrix} x_1 \\ x_2 \\ x_3 \end{pmatrix}, \text{ and } k = \begin{pmatrix} k_1 \\ k_2 \\ k_3 \end{pmatrix}$$

the more complicated expression:

$$a_{11}(a_{22}a_{33} - a_{23}a_{32}) - a_{12}(a_{21}a_{33} - a_{23}a_{31}) + a_{13}(a_{21}a_{32} - a_{22}a_{31})$$

arises. As before we call this the determinant of A and denote it by $|A|$. We note that we can write this determinant in terms of the determinants of 2×2 matrices as follows.

$$\begin{vmatrix} a_{11} & a_{12} & a_{13} \\ a_{21} & a_{22} & a_{23} \\ a_{31} & a_{32} & a_{33} \end{vmatrix} = a_{11}\begin{vmatrix} a_{22} & a_{23} \\ a_{32} & a_{33} \end{vmatrix} - a_{12}\begin{vmatrix} a_{21} & a_{23} \\ a_{31} & a_{33} \end{vmatrix} + a_{13}\begin{vmatrix} a_{21} & a_{22} \\ a_{31} & a_{32} \end{vmatrix}.$$

This suggests a way of defining determinants of arbitrary $n \times n$ matrices. However it would be tedious and not easy to deduce the properties of determinants in this way. Yet, just as with matrices, it is our ability to calculate with our shorthand notations that make them so important. For this reason we turn to a more convenient way of defining determinants. This depends on the idea of a permutation, to which we now turn.

4.3 PERMUTATIONS

A permutation is a one-one function of a finite set S onto itself. There is no loss in generality in labelling the elements of the set with the natural numbers 1 to n. Thus we take the set S to be

$$\{1,2,3,4, \ldots\ldots, n\}.$$

If f is a permutation which maps the number

$$m \text{ to } f(m), m = 1,2,3,4, \ldots\ldots, n,$$

we often write

$$\begin{pmatrix} 1 & 2 & 3 \ldots\ldots\ldots\ldots & m \\ f(1) & f(2) & f(3) \ldots\ldots\ldots & f(m) \end{pmatrix} .$$

Let f and g be two permutations

$$S \xrightarrow{f} S \text{ and } S \xrightarrow{g} S.$$

We define the product gf of the two permutations to be the composition of f and g as functions in the sense of section 2.11. We apply f first then g.

Thus $\qquad (gf)(m) = g(f(m)).$

In terms of the idea of a product of permutations we can express a given permutation in two ways, both convenient for certain purposes.

First we need the idea of a **cycle**. This is the special permutation

$$\begin{pmatrix} a_1 & a_2 \ldots\ldots\ldots & a_r \\ a_2 & a_3 \ldots\ldots\ldots & a_1 \end{pmatrix}, \ a_i \in S .$$

For example:

$$\begin{pmatrix} 1 & 2 & 3 \\ 2 & 3 & 1 \end{pmatrix}, \qquad \begin{pmatrix} 2 & 7 & 3 \\ 7 & 3 & 2 \end{pmatrix} .$$

We leave out the symbols mapped onto themselves.
It is usual to write a cycle in the abbreviated form:

$$f = (a_1, a_2, a_3, \ldots\ldots, a_r),$$

where $\quad f(a_i) = a_{i+1}, i = 1,2, \ldots\ldots, r-1, \text{ and } f(a_r) = a_1.$

Here r is called the **length** of the cycle.
Every permutation can be written as a product of disjoint cycles. Rather than get involved in the complexities of the general case, we will illustrate the procedure by an example which easily extends to any permutation.

Let $$f = \begin{pmatrix} 1 & 2 & 3 & 4 & 5 & 6 & 7 & 8 \\ 3 & 7 & 6 & 5 & 8 & 1 & 2 & 4 \end{pmatrix}.$$

Start with the 1 and write the first cycle as

$$(1, f(1), f(f(1)), \ldots\ldots, f^r(1) = 1),$$

where f^r means f multiplied by itself r times.
Next look for a symbol that has not yet appeared in the cycle or cycles obtained so far. Let this be i.
The next cycle is

$$(i, f(i), f(f(i)), \ldots\ldots, f^t(i) = i).$$

Repeat this procedure until all symbols have appeared in a cycle.
In our example this gives:

$$f = (1, 3, 6)(2, 7)(4, 5, 8)$$

If we note that

$$(a_1, a_2, a_3, \ldots\ldots, a_r) = (a_1, a_r)(a_1, a_{r-1}) \ldots\ldots (a_1, a_2)$$

then we can see that every permutation can be written as a product of cycles of length two, **not necessarily disjoint**. Cycles of length two are usually called **transpositions**.

For example:

$$f = \begin{pmatrix} 1 & 2 & 3 & 4 & 5 & 6 \\ 4 & 3 & 5 & 6 & 2 & 1 \end{pmatrix} \text{ can be written } (1, 4, 6)(2, 3, 5)$$

$$\text{or} \quad (1, 6)(1, 4)(2, 5)(2, 3).$$

Since the composition of functions is an associative binary operation it follows that the product of permutations is also an associative binary operation on the set of all permutations of the set S. This set of permutations on S is denoted by S_n and has $n!$ members when

$$S = \{1, 2, 3, 4, \ldots\ldots, n\}.$$

We also note that the permutation

$$\begin{pmatrix} 1 & 2 & 3 & 4 & \cdots\cdots & n \\ 1 & 2 & 3 & 4 & \cdots\cdots & n \end{pmatrix}$$

acts as the identity element for our product and that

$$\begin{pmatrix} a_1 & a_2 & a_3 & \cdots\cdots & a_n \\ b_1 & b_2 & b_3 & \cdots\cdots & b_n \end{pmatrix}$$

is the multiplicative inverse to

$$\begin{pmatrix} b_1 & b_2 & b_3 & \cdots\cdots & b_n \\ a_1 & a_2 & a_3 & \cdots\cdots & a_n \end{pmatrix} .$$

A set like S_n with a product having the above properties is called a **group**. This concept is vital for the study of the fundamental notion of symmetry. For future reference we give the following definition.

DEFINITION 4.3.1 A **group** is a set G together with a binary operation

$$G \times G \longrightarrow G, \text{ where } (a, b) \longrightarrow ab,$$

such that

(1) $a(bc) = (ab)c$ for all $a, b, c \in G$

(2) there exists an element $e \in G$ satisfying

$$ea = a = ae \text{ for all } a \in G$$

(3) associated with each element $a \in G$ there is an inverse element a^{-1} with the property that

$$a\ a^{-1} = e = a^{-1}\ a.$$

If also

(4) $ab = ba$ for all $a, b \in G$,
then the group is said to be **abelian** or **commutative**.

Note that the identity element e in (2) is **unique** and that for a given a the inverse a^{-1} defined by (3) is **unique**. To see this, let e' be a second identity.

Then $e = e\,e' = e'$. Let a' and a'' both be inverses of a.

Then $\quad a' = a'e = a'(aa'') = (a'a)a'' = ea'' = a''$.

In terms of the concept of the group we can describe a field very neatly as an abelian group under addition (+), and, leaving out the zero, which is the identity element of the additive group of the field, an abelian group under multiplication; the two operations being linked by the distributive rules. In a non-commutative field the multiplicative group of non-zero elements is non-abelian.

A useful notion is the order of an element of a group. We denote $aaaaaa\ldots a$ to t factors by a^t. In a finite group (i.e. a group with a finite number of elements) the sequence $a, a^2, a^3, \ldots\ldots$ must repeat. Suppose $a^r = a^s$. Then if $r > s$, we have

$$a^{r-s} = a^r(a^{-1})^s = a^s(a^{-1})^s = e.$$

The smallest positive integer t so that $a^t = e$ is called the **order of a**. Not to be confused with this is the **order of the group** which is just the number of elements in the group.

EXAMPLE 4.3.1 In the finite group S_3 the order of (1,2,3) is 3. In S_8 the order of (1,3,6) is 3, of (2,7) is 2 and of (4,5,8) is 3. The order of (1,3,6)(2,7)(4,5,8) is 6. These illustrate the general results that the order of a cycle of length r is r and the order of a product of **disjoint** cycles is the lowest common multiple of the orders of the individual cycles. The reader is invited to prove these results.

In order to apply permutations to the definition of determinants, we have to associate a sign (+ or –) with each permutation. To this end we define a function:

$$\sigma : S_n \longrightarrow \mathbf{Q} \text{ by}$$

$$\sigma(f) = \Pi\left(\frac{a_i - a_j}{b_i - b_j}\right), \text{ where } f = \begin{pmatrix} a_1 & a_2 & a_3 \ldots\ldots & a_n \\ b_1 & b_2 & b_3 \ldots\ldots & b_n \end{pmatrix}$$

and the product is taken over all subsets $\{i, j\}$ of $\{1,2,3,4\ldots\ldots, n\}$. **Note** that our definition is **not** dependent on the order in which we write the columns

$\begin{pmatrix} a_i \\ b_i \end{pmatrix}$. For example if $f = \begin{pmatrix} 1 & 2 & 3 & 4 \\ 2 & 3 & 4 & 1 \end{pmatrix}$

then $$\sigma(f) = \left(\frac{1-2}{2-3}\right)\left(\frac{1-3}{2-4}\right)\left(\frac{1-4}{2-1}\right)\left(\frac{2-3}{3-4}\right)\left(\frac{2-4}{3-1}\right)\left(\frac{3-4}{4-1}\right)$$

$$= -1$$

Now if $f = \begin{pmatrix} a_1 & a_2 \ldots\ldots a_n \\ b_1 & b_2 \ldots\ldots b_n \end{pmatrix}$ and $g = \begin{pmatrix} b_1 & b_2 & b_3 \ldots\ldots b_n \\ c_1 & c_2 & c_3 \ldots\ldots c_n \end{pmatrix}$

then $gf = \begin{pmatrix} a_1 & a_2 \ldots\ldots a_n \\ c_1 & c_2 \ldots\ldots c_n \end{pmatrix}$ and

$$\sigma(gf) = \Pi \left(\frac{a_i - a_j}{c_i - c_j}\right) = \Pi \left(\frac{b_i - b_j}{c_i - c_j}\right) \quad \Pi \left(\frac{a_i - a_j}{b_i - b_j}\right).$$

Thus $\sigma(gf) = \sigma(g) \quad \sigma(f)$.

DEFINITION 4.3.2 Let G and H be two groups. Let f be a function from G into H. Then if

$$f(g_1 g_2) = f(g_1) f(g_2) \text{ for all } g_1, g_2 \in G$$

we call f a homomorphism of G into H.

In terms of our definition we have proved that σ is a homomorphism of the group S_n into the multiplicative group of the field **Q**.

In fact the image of σ is much smaller than **Q**. To show this we find $\sigma(t)$, where t denotes a transposition. There is no loss in generality in supposing that

$$t = \begin{pmatrix} a_1 & a_2 & a_3 & a_4 \ldots\ldots a_n \\ a_2 & a_1 & a_3 & a_4 \ldots\ldots a_n \end{pmatrix}$$

Then $\sigma(t) = \dfrac{a_1 - a_2}{a_2 - a_1} \quad \dfrac{a_1 - a_3}{a_2 - a_3} \quad \dfrac{a_1 - a_4}{a_2 - a_4} \ldots\ldots \dfrac{a_1 - a_n}{a_2 - a_n} \quad \dfrac{a_2 - a_3}{a_1 - a_3} \ldots\ldots$

$$\frac{a_2 - a_n}{a_1 - a_n} \quad \prod_{\substack{i \neq 1,2 \\ j \neq 1,2}} \left(\frac{a_i - a_j}{a_i - a_j}\right).$$

Then $\sigma(t) = \dfrac{a_1 - a_2}{a_2 - a_1} = -1.$

Now let f be any permutation. As we have mentioned, f can be written as the product of transpositions. In this form let

$$f = t_1 t_2 t_3 t_4 \ldots\ldots t_m$$

where the t_i are transpositions.

For each t_i, $\sigma(t_i) = -1$ and, since σ is a homomorphism,

$$\sigma(f) = \sigma(t_1)\ \sigma(t_2)\ \ \sigma(t_3)\ldots\ldots\ldots\sigma(t_m).$$

$$\text{Thus} \quad \sigma(f) = (-1)^m \begin{array}{l} = +1 \text{ if } m \text{ is even,} \\ = -1 \text{ if } m \text{ is odd.} \end{array}$$

In this way we associate with each permutation a positive or negative sign. Equivalently we refer to a permutation as being even or odd. As seen above, to find $\sigma(f)$ we write f as a product of transpositions. However, if we note that a cycle of length r can be written as a product of $(r-1)$ transpositions, it suffices to express the permutation f as a product of disjoint cycles.

EXAMPLE 4.3.2

$$\text{Let } f = \begin{pmatrix} 1 & 2 & 3 & 4 & 5 & 6 & 7 & 8 & 9 \\ 3 & 4 & 7 & 6 & 9 & 8 & 1 & 2 & 5 \end{pmatrix}$$

$$\begin{aligned} \text{Then } \sigma(f) &= \sigma[(1,3,7)\,(2,4,6,8)\,(5,9)] \\ &= \sigma(1,3,7)\quad \sigma(2,4,6,8)\quad \sigma(5,9) \\ &= (-1)^{3-1} \cdot\ (-1)^{4-1}\quad (-1)^{2-1} \quad = (-1)^2 \cdot\ (-1)^3 \cdot\ (-1) \\ &= 1 \cdot (-1) \cdot (-1) \quad = \quad 1. \end{aligned}$$

Note that interchanging two symbols in a permutation i.e. multiplying by a transposition changes the sign or parity of a permutation.

4.4 DEFINITION AND PROPERTIES OF $n \times n$ DETERMINANT

We are now in a position to give the following

DEFINITION 4.4.1 Let A be an $n \times n$ matrix with elements a_{ij} from a field F. The determinant of A, denoted by $|A|$ is defined to be:

$$\Sigma\,\sigma\begin{pmatrix}1 & 2 & 3 \ldots\ldots n\\ j_1 & j_2 & j_3 \ldots\ldots j_n\end{pmatrix} \quad a_{1j_1}\ a_{2j_2}\ a_{3j_3}\ a_{4j_4}\ a_{5j_5}\ \cdots\cdots\ a_{nj_n},$$

where the sum Σ is taken over **all** $n!$ permutations

$$\begin{pmatrix}1 & 2 & 3 \ldots\ldots n\\ j_1 & j_2 & j_3 \ldots\ldots j_n\end{pmatrix} \quad \text{of the set} \quad \{1, 2, 3, 4, \ldots\ldots, n\}.$$

EXAMPLE 4.4.1 To illustrate this definition we use it to calculate determinants of orders 2 and 3. (In general the determinant of an $n \times n$ matrix is said to be of order n). At the same time this will verify that our definition does agree with that given before for these special cases.

$$\begin{vmatrix} a_{11} & a_{12}\\ a_{21} & a_{22}\end{vmatrix} = \sigma\begin{pmatrix}12\\12\end{pmatrix} a_{11}\ a_{22} + \sigma\begin{pmatrix}12\\21\end{pmatrix} a_{12}\ a_{21}.$$

$$= (+1)\ a_{11}\ a_{22} + (-1)\ a_{12}\ a_{21}$$

$$= a_{11}\ a_{22} - a_{12}\ a_{21}$$

$$\begin{vmatrix} a_{11} & a_{12} & a_{13}\\ a_{21} & a_{22} & a_{23}\\ a_{31} & a_{32} & a_{33}\end{vmatrix} = \sigma\begin{pmatrix}1 & 2 & 3\\ 1 & 2 & 3\end{pmatrix} a_{11}\ a_{22}\ a_{33} + \sigma\begin{pmatrix}1 & 2 & 3\\ 2 & 1 & 3\end{pmatrix} a_{12}\ a_{21}\ a$$

$$+ \sigma\begin{pmatrix}1 & 2 & 3\\ 3 & 2 & 1\end{pmatrix} a_{13}\ a_{22}\ a_{31} + \sigma\begin{pmatrix}1 & 2 & 3\\ 1 & 3 & 2\end{pmatrix} a_{11}\ a_{23}$$

$$+ \sigma\begin{pmatrix}1 & 2 & 3\\ 2 & 3 & 1\end{pmatrix} a_{12}\ a_{23}\ a_{31} + \sigma\begin{pmatrix}1 & 2 & 3\\ 3 & 1 & 2\end{pmatrix} a_{13}\ a_{21}$$

Now $\sigma\begin{pmatrix}1 & 2 & 3\\ 1 & 2 & 3\end{pmatrix} = 1,\ \sigma\begin{pmatrix}1 & 2 & 3\\ 2 & 1 & 3\end{pmatrix} = \sigma(12) = -1,$

$\sigma\begin{pmatrix}1 & 2 & 3\\ 3 & 2 & 1\end{pmatrix} = \sigma(13) = -1,\ \sigma\begin{pmatrix}1 & 2 & 3\\ 1 & 3 & 2\end{pmatrix} = \sigma(23) = -1,$

$$\sigma\begin{pmatrix} 1 & 2 & 3 \\ 2 & 3 & 1 \end{pmatrix} = \sigma(1\ 2\ 3) = (-1)^2 = 1,$$

$$\sigma\begin{pmatrix} 1 & 2 & 3 \\ 3 & 1 & 2 \end{pmatrix} = \sigma(1\ 3\ 2) = (-1)^2 = 1.$$

Thus we have:

$$\begin{vmatrix} a_{11} & a_{12} & a_{13} \\ a_{21} & a_{22} & a_{23} \\ a_{31} & a_{32} & a_{33} \end{vmatrix} = a_{11}\ a_{22}\ a_{33} - a_{12}\ a_{21}\ a_{33} +$$

$$- a_{13}\ a_{22}\ a_{31} - a_{11}\ a_{23}\ a_{32} +$$

$$+ a_{12}\ a_{23}\ a_{31} + a_{13}\ a_{21}\ a_{32}$$

$$= a_{11}\begin{vmatrix} a_{22} & a_{23} \\ a_{32} & a_{33} \end{vmatrix} - a_{12}\begin{vmatrix} a_{21} & a_{23} \\ a_{31} & a_{33} \end{vmatrix} + a_{13}\begin{vmatrix} a_{21} & a_{22} \\ a_{31} & a_{32} \end{vmatrix}.$$

These are in agreement with our earlier work.

In order to work with determinants, it is convenient to have available certain elementary properties. We list these below as:

THEOREM 4.4.1

(1) If every element of a row (column) of the $n \times n$ matrix A is zero then $|A| = 0$.

(2) $|A'| = |A|$.

(3) If every element of a row (column) of A is multiplied by a scalar $k \in F$, then $|A|$ is multiplied by k. If every element of a row (column) of A has k as a factor, then k may be factored out of $|A|$.

(4) If B is obtained from A by interchanging any two rows (columns) of A, then $|A| = -|B|$.

(5) If two rows (columns) of A are identical, then $|A| = 0$. We assume the field F does not contain $\mathbf{Z}_2$.

(6) If the ith row (column) of A is the sum of p rows (columns) of elements, then $|A|$ = the sum of p determinants obtained by taking each summand of the ith row (column) separately.

(7) If B is obtained from A by adding k times the ith row (column) of A to the jth row (column) of A, then $|B| = |A|$.

(Here the rows (columns) are added as vectors e.g.

$$(a_{11}, a_{12}, a_{13}) + (a_{31}, a_{32}, a_{33}) = (a_{11} + a_{31}, a_{12} + a_{32}, a_{13} + a_{33})).$$

PROOF

(1) Each term in $|A|$ contains one element from the ith row (column) for each i. Thus if the ith row (column) is zero then each term of $|A|$ is zero. Hence $|A|$ is zero.

(2) $$|A| = \Sigma\sigma\begin{pmatrix}1 & 2 & 3 \ldots & n\\ j_1 & j_2 & j_3 \ldots & j_n\end{pmatrix} a_{1j_1}\ a_{2j_2}\ a_{3j_3}\ a_{4j_4} \ldots\ldots a_{nj_n}.$$

We rearrange the order of the a_{ij_r} so that the second suffices run from 1 to n in order. Thus

$$|A| = \Sigma\sigma\begin{pmatrix}1 & 2 & 3 \ldots & n\\ j_1 & j_2 & j_3 \ldots & j_n\end{pmatrix} a_{f(1)1}\ a_{f(2)2}\ a_{f(3)3}\ a_{f(4)4} \ldots\ldots a_{f(n)n}$$

The permutation f is the **inverse** of

$$\begin{pmatrix}1 & 2 & 3 \ldots\ldots & n\\ j_1 & j_2 & j_3 \ldots\ldots & j_n\end{pmatrix}$$

If $\begin{pmatrix}1 & 2 & 3 \ldots\ldots & n\\ j_1 & j_2 & j_3 \ldots\ldots & j_n\end{pmatrix} = t_1\, t_2 \ldots\ldots t_s$, where the t_i are transpositions,

then $$f = t_s\, t_{s-1} \ldots\ldots t_1.$$

Thus $$\sigma(f) = \sigma\begin{pmatrix}1 & 2 & 3 \ldots\ldots & n\\ j_1 & j_2 & j_3 \ldots\ldots & j_n\end{pmatrix} = (-1)^s.$$

Hence $$|A| = \Sigma\sigma(f)\ a_{f(1)1}\ a_{f(2)2}\ a_{f(3)3}\ a_{f(4)4} \ldots\ldots a_{f(n}$$

$$= \Sigma\sigma(f)\ a'_{1f(1)}\ a'_{2f(2)}\ a'_{3f(3)} \ldots\ldots a'_{nf(n)} =$$

$$= |A'|, \text{ where } a'_{rt} = a_{tr}.$$

(The reader is advised to work a particular case of the above in order to see how the rearrangement affects the permutation of the suffices).

(3) Since each term of $|B|$, where B is obtained from A by multiplying the ith row by k, has one element from its ith row, we have:

$$|B| = \Sigma \sigma \begin{pmatrix} 1 & 2 & 3 \ldots\ldots & n \\ j_1 & j_2 & j_3 \ldots\ldots & j_n \end{pmatrix} a_{1j_1}\ a_{2j_2}\ a_{3j_3}\ a_{4j_4} \ldots (ka_{ij_i}) \ldots . a_{nj_n} .$$

$$= k\, \Sigma \sigma \begin{pmatrix} 1 & 2 & 3 \ldots\ldots & n \\ j_1 & j_2 & j_3 \ldots\ldots & j_n \end{pmatrix} a_{1j_1}\ a_{2j_2}\ a_{3j_3}\ a_{4j_4} \ldots\ldots a_{ij_i} \ldots\ldots a_{nj_n} .$$

$$= k|A| .$$

Similarly for columns, in fact part (2) above shows that any result proved for rows also holds for columns. Hence in the remaining parts we shall prove only for rows.

(4) Interchanging any two rows alters $\begin{pmatrix} 1 & 2 & 3 \ldots\ldots & n \\ j_1 & j_2 & j_3 \ldots\ldots & j_n \end{pmatrix}$

by a transposition. Hence

$$\sigma \begin{pmatrix} 1 & 2 \ldots\ldots & n \\ j_1 & j_2 \ldots\ldots & j_n \end{pmatrix} = -\sigma \begin{pmatrix} 1 & 2 & 3 \ldots\ldots & n \\ j_1' & j_2' & j_3' \ldots\ldots & j_n' \end{pmatrix}$$

for **all** $\begin{pmatrix} 1 & 2 \ldots\ldots & n \\ j_1 & j_2 \ldots\ldots & j_n \end{pmatrix}$, where $\begin{pmatrix} 1 & 2 \ldots\ldots & n \\ j_1' & j_2' \ldots\ldots & j_n' \end{pmatrix}$

is the new arrangement brought about by the interchange of rows.

Thus each term of $|A|$ changes sign.

Hence $\quad |B| = -|A|$.

(5) Interchange the identical rows. By (4) $|A| = -|A|$.
Then provided F does not contain $\mathbf{Z}_2$, we may deduce that $|A| = 0$.

(6) If the ith row of A is $(d_{i1} + b_{i1}, d_{i2} + b_{i2}, \ldots\ldots, d_{in} + b_{in})$

$$\text{then} \quad |A| = \Sigma\sigma\begin{pmatrix}1 & 2 \ldots\ldots & n\\ j_1 & j_2 \ldots\ldots & j_n\end{pmatrix} a_{1j_1} \ldots (d_{ij_i} + b_{ij_i}) \ldots a_{nj_n}$$

$$= \Sigma\sigma\begin{pmatrix}1 & 2 \ldots\ldots & n\\ j_1 & j_2 \ldots\ldots & j_n\end{pmatrix} a_{1j_1} \ldots d_{ij_i} \ldots\ldots a_{nj_n} +$$

$$+ \Sigma\sigma\begin{pmatrix}1 & 2 \ldots\ldots & n\\ j_1 & j_2 \ldots\ldots & j_n\end{pmatrix} a_{1j_1} \ldots b_{ij_i} \ldots\ldots a_{nj_n}.$$

Hence $|A| = |D| + |B|$.

This proves the required result for $p = 2$. The same method proves the general case.

(7) $$|B| = \begin{vmatrix} a_{11} & a_{12} \ldots\ldots & a_{1n}\\ \ldots & \ldots & \ldots\\ a_{j1} + ka_{i1} & a_{j2} + ka_{i2} \ldots\ldots & a_{jn} + ka_{in}\\ \ldots & \ldots & \ldots \end{vmatrix} =$$

$$= \begin{vmatrix} a_{11} & a_{12} & a_{1n}\\ \ldots & \ldots & \ldots\\ a_{j1} & a_{j2} & a_{jn}\\ \ldots & \ldots & \ldots\\ a_{n1} & a_{n2} & a_{nn}\end{vmatrix} + \begin{vmatrix} a_{11} & a_{12} & a_{1n}\\ \ldots & \ldots & \ldots\\ ka_{i1} & ka_{i2} & ka_{in}\\ \ldots & \ldots & \ldots\\ a_{n1} & a_{n2} & a_{nn}\end{vmatrix}$$

by (6). Then by (3) we have:

$$|B| = |A| + k\begin{vmatrix} a_{11} \ldots\ldots a_{1n}\\ \ldots\ldots\ldots\\ a_{i1} \ldots\ldots a_{in}\\ \ldots\ldots\ldots\\ a_{i1} \ldots\ldots a_{in}\\ \ldots\ldots\ldots \end{vmatrix} \begin{matrix} \\ \\ \leftarrow j\text{th row.}\\ \\ \leftarrow i\text{th row.}\\ \\ \end{matrix}$$

Now the last determinant is zero by (5).

Hence finally: $|B| = |A|$.

EXAMPLE 4.4.2 Using the above properties we evaluate the determinant:

$$\begin{vmatrix} 1 & 2 & 3 \\ 2 & 4 & 2 \\ 3 & 5 & 1 \end{vmatrix}$$

By (7) we take twice the first row from the third row. We then use (2) and expand by the first row according to Example 4.4.1. The results are given below.

$$\begin{vmatrix} 1 & 2 & 3 \\ 2 & 4 & 2 \\ 3 & 5 & 1 \end{vmatrix} = \begin{vmatrix} 1 & 2 & 3 \\ 0 & 0 & -4 \\ 0 & -1 & -8 \end{vmatrix} = \begin{vmatrix} 1 & 0 & 0 \\ 2 & 0 & -1 \\ 3 & -4 & -8 \end{vmatrix} = 1 \begin{vmatrix} 0 & -1 \\ -4 & -8 \end{vmatrix} = -4.$$

4.5 MINORS AND COFACTORS

Let A be an $m \times n$ matrix. Cross out r rows and s columns. Form the $r \times s$ matrix whose elements are the elements of A at the intersection points. We call this matrix a **minor** of A. In particular if A is an $n \times n$ matrix and we cross out $n - 1$ rows and $n - 1$ columns, then the intersection points give us an $(n - 1) \times (n - 1)$ matrix. We denote this matrix by M_{ij}, where i is the row and j is the column NOT crossed out. M_{ij} is called a **first minor** of A and $|M_{ij}|$ is called a **first minor** of the corresponding determinant $|A|$. The signed minor $(-1)^{i+j}|M_{ij}|$ is called the **cofactor** of the element a_{ij} of

$$A = \begin{pmatrix} a_{11} & a_{12} \cdot \cdot \cdot \cdot a_{1n} \\ a_{21} & a_{22} \cdot \cdot \cdot \cdot a_{2n} \\ \cdots & \cdots \\ a_{n1} & a_{n2} \cdot \cdot \cdot a_{nn} \end{pmatrix}.$$

We usually symbolise the cofactor of a_{ij} by A_{ij}.

The idea of the cofactor provides us with a convenient notation for describing the expansion of a determinant. This is the content of

THEOREM 4.5.1 The value of $|A|$ is the sum

$$a_{i1} A_{i1} + a_{i2} A_{i2} + a_{i3} A_{i3} + \ldots\ldots + a_{ij} A_{ij} + \ldots\ldots + a_{in} A_{in}.$$

This is called the expansion of $|A|$ by its ith row. A similar expression is true for columns. Thus

$$|A| = a_{1j}A_{1j} + a_{2j}A_{2j} + a_{3j}A_{3j} + \ldots\ldots + a_{ij}A_{ij} + \ldots\ldots + a_{nj}A_{nj}$$

is the expansion of $|A|$ by its jth column.

PROOF

$|A|$ = terms involving a_{i1} + terms involving a_{i2} + terms involving a_{i3} + ... + + terms involving a_{in}. We consider the terms involving a_{ij}. Starting with A we bring the ith row to the position of the top row and the jth column to the position of the first column by a sequence of interchanges of rows and of columns, without altering the order of the remaining rows and columns. We call the resulting matrix D. We have:

$$D = \begin{vmatrix} a_{ij} & a_{i1} \cdots\cdots & a_{i,j-1} & a_{i,j+1} \cdots & a_{in} \\ a_{1j} & a_{11} \cdots\cdots & a_{1,j-1} & a_{1,j+1} \cdots & a_{1n} \\ a_{2j} & a_{21} \cdots\cdots & a_{2,j-1} & a_{2,j+1} \cdots & a_{2n} \\ \cdot & \cdot & \cdot & \cdot & \cdot \\ \cdot & \cdot & \cdot & \cdot & \cdot \\ \cdot & \cdot & \cdot & \cdot & \cdot \\ \cdot & \cdot & \cdot & \cdot & \cdot \\ a_{nj} & a_{n1} \cdots\cdots & a_{n,j-1} & a_{n,j+1} \cdots & a_{nn} \end{vmatrix} =$$

$$= \begin{vmatrix} a_{ij} & a_{i1} \cdots\cdots a_{i,j-1} \quad a_{i,j+1} \cdots a_{in} \\ a_{1j} & \\ a_{2j} & \\ \cdot & \\ \cdot & \\ \cdot & M_{ij} \\ \cdot & \\ a_{nj} & \end{vmatrix}$$

In the transition from A to D we have used $(i-1)+(j-1)$ interchanges of rows and columns.

Thus $\quad |D| = (-1)^{i+j-2}\,|A| = (-1)^{i+j}\,|A|$.

Now the term involving a_{ij} in $|D|$ is $a_{ij}\,|M_{ij}|$.

Hence the term involving a_{ij} in $|A|$ is $a_{ij}\,(-1)^{i+j}\,|M_{ij}| = a_{ij}\,A_{ij}$.

Hence $|A| = \sum_{j=1}^{n} a_{ij}\,A_{ij}$ as required.

By theorem 4.4.1 part (2), the required result now holds for columns. This completes the proof of theorem 4.5.1.

EXAMPLE 4.5.1

We expand $|A| = \begin{vmatrix} 1 & -3 & 2 \\ 0 & 1 & -4 \\ 2 & 0 & 1 \end{vmatrix}$

by its second row. We have:

$$|A| = 0 \cdot A_{21} + 1 \cdot A_{22} + (-4) \cdot A_{23}$$

$$= 0 + (-1)^{2+2} \begin{vmatrix} 1 & 2 \\ 2 & 1 \end{vmatrix} + (-4) \cdot (-1)^{2+3} \begin{vmatrix} 1 & -3 \\ 2 & 0 \end{vmatrix}$$

$$= (1-4) + (-4)(-1)\,6 = -3 + 24 = 21.$$

The expansion of A by its third column takes the form:

$$|A| = 2 \cdot A_{13} + (-4) \cdot A_{23} + 1 \cdot A_{33}$$

$$= 2 \cdot (-1)^{1+3} \begin{vmatrix} 0 & 1 \\ 2 & 0 \end{vmatrix} + (-4) \cdot (-1)^{2+3} \cdot \begin{vmatrix} 1 & -3 \\ 2 & 0 \end{vmatrix} +$$

$$+ \quad 1 \cdot (-1)^{3+3} \cdot \begin{vmatrix} 1 & -3 \\ 0 & 1 \end{vmatrix}$$

$$= 2(-2) + (-4)(-1)\,6 + 1 \cdot 1$$

$$= -4 + 24 + 1 = 21.$$

Let A and B be two $n \times n$ matrices over the field F i.e. their elements come from F. The product AB is also an $n \times n$ matrix. Thus $|AB|$ exists. It is natural to wonder if there is any simple relation connecting $|AB|$ with $|A||B|$. However, before studying this it is convenient to state and prove a preliminary result.

THEOREM 4.5.2

Let A and B be two $n \times n$ matrices over a field F. Let O denote the $n \times n$ zero matrix.

Then $|A||B| = \begin{vmatrix} A & O \\ C & B \end{vmatrix}$, where the latter is a $2n \times 2n$ determinant and C is any $n \times n$ matrix.

PROOF

$$\text{Let} \quad D = \begin{pmatrix} A & O \\ C & B \end{pmatrix} = (d_{ij}), \begin{array}{l} i = 1, 2, \ldots\ldots, 2n \\ j = 1, 2, \ldots\ldots, 2n \end{array}.$$

In more detail we have:

$$D = \begin{pmatrix} a_{11} & a_{12} \cdots\cdots a_{1n} & 0 & 0 \cdots\cdots 0 \\ a_{21} & a_{22} \cdots\cdots a_{2n} & 0 & 0 \cdots\cdots 0 \\ \cdot & \cdot \qquad\quad \cdot & \cdot & \cdot \qquad\quad \cdot \\ \cdot & \cdot \qquad\quad \cdot & \cdot & \cdot \qquad\quad \cdot \\ \cdot & \cdot \qquad\quad \cdot & \cdot & \cdot \qquad\quad \cdot \\ a_{n1} & a_{n2} \cdots\cdots a_{nn} & 0 & 0 \cdots\cdots 0 \\ c_{11} & c_{12} \cdots\cdots c_{1n} & b_{11} & b_{12} \cdots b_{1n} \\ \cdot & \cdot \qquad\quad \cdot & \cdot & \cdot \qquad\quad \cdot \\ \cdot & \cdot \qquad\quad \cdot & \cdot & \cdot \qquad\quad \cdot \\ \cdot & \cdot \qquad\quad \cdot & \cdot & \cdot \qquad\quad \cdot \\ c_{n1} & c_{n2} \cdots\cdots c_{nn} & b_{n1} & b_{n2} \cdots b_{nn} \end{pmatrix}$$

By definition

$$D = \Sigma \sigma \begin{pmatrix} 1 & 2 \ldots & 2n \\ j_1 & j_2 \ldots & j_{2n} \end{pmatrix} \quad d_{1j_1}\ d_{2j_2}\ d_{3j_3} \cdots\cdots d_{2n, j_{2n}}.$$

Because of the block of zeros in the top right-hand corner of D, $\{j_1, j_2 \ldots . j_n\}$ is just a rearrangement of $\{1, 2, \ldots\ldots, n\}$ and $\{j_{n+1}, j_{n+2}, \ldots\ldots, j_{2n}\}$ is a rearrangement of $\{n+1, n+2, \ldots\ldots, 2n\}$.

Thus $\sigma\begin{pmatrix} 1 & 2 \ldots & 2n \\ j_1 & j_2 \ldots & j_{2n} \end{pmatrix} = \sigma\begin{pmatrix} 1 & 2 \ldots n & \cdot & n+1 & n+2 \ldots\ldots 2n \\ j_1 & j_2 \ldots j_n & \cdot & j_{n+1} & j_{n+2} \ldots\ldots j_{2n} \end{pmatrix}$

$$= \sigma\ (t_1\, t_2 \ldots t_r \cdot k_1\, k_2 \ldots k_s) = \sigma\ (t_1\, t_2 \ldots t_r) \cdot \sigma\, (k_1\, k_2 \ldots k_s)$$

where $t_1, t_2, \ldots\ldots, t_r$ are transpositions involving **only** the numbers 1 to n and $k_1, k_2, \ldots\ldots, k_s$ are transpositions involving just $n + 1$ to $2n$.

Hence $\sigma\begin{pmatrix} 1 & 2 \ldots & 2n \\ j_1 & j_2 \ldots & j_{2n} \end{pmatrix} = \sigma\begin{pmatrix} 1 & 2 \ldots & n \\ j_1 & j_2 \ldots & j_n \end{pmatrix} \sigma\begin{pmatrix} n+1 & n+2 \ldots & 2n \\ j_{n+1} & j_{n+2} \ldots & j_{2n} \end{pmatrix}$

Thus $|D| = \Sigma\, \sigma\begin{pmatrix} 1 & 2 \ldots & n \\ j_1 & j_2 \ldots & j_n \end{pmatrix} \sigma\begin{pmatrix} n+1 \ldots & 2n \\ j_{n+1} \ldots & j_{2n} \end{pmatrix} d_{1j_1}\, d_{2j_2}\, d_{3j_3} \ldots . d_{2n,j_{2n}}$

$$= \Sigma\, \sigma\begin{pmatrix} 1 & 2 \ldots & n \\ j_1 & j_2 \ldots & j_n \end{pmatrix} d_{1j_1}\, d_{2j_2} \ldots . d_{nj_n} \cdot \sigma\begin{pmatrix} n+1, \ldots, 2n \\ j_{n+1}, \ldots, j_{2n} \end{pmatrix}$$

$$d_{n+1,j_{n+1}} \ldots\ldots d_{2n,j_{2n}}$$

$$= \Sigma\, \sigma\begin{pmatrix} 1 & 2 \ldots & n \\ j_1 & j_2 \ldots & j_n \end{pmatrix} d_{1j_1}\, d_{2j_2} \ldots\ldots d_{nj_n}\ \Sigma\, \sigma\begin{pmatrix} n+1 \ldots & 2n \\ j_{n+1} \ldots & j_{2n} \end{pmatrix}$$

$$d_{n+1,j_{n+1}} \ldots\ldots d_{2n,j_{2n}}$$

$$= \Sigma\, \sigma\begin{pmatrix} 1 \ldots & n \\ j_1 \ldots & j_n \end{pmatrix} a_{1j_1}\, a_{2j_2} \ldots\ldots a_{nj_n}\ \Sigma\, \sigma\begin{pmatrix} 1 \ldots & n \\ l_1 \ldots & l_n \end{pmatrix}$$

$$b_{1l_1}\, b_{2l_2} \ldots\ldots b_{nl_n}$$

$= |A| \cdot |B|$, as required.

The connection between $|AB|$ and $|A||B|$ has the form we would like. Formally we have:

THEOREM 4.5.3

$$|AB| = |A||B|, \text{ where } A \text{ and } B \text{ are } n \times n \text{ matrices.}$$

PROOF

Let $A = (a_{ij})$ and $B = (b_{ij})$.

Let $C = AB$, where $c_{ij} = \sum_{k=1}^{n} a_{ik} b_{kj}$.

By theorem 4.5.2 we have:

$$|A||B| = \begin{vmatrix} a_{11} & a_{12} \cdots & a_{1n} & 0 & 0 \cdots & 0 \\ a_{21} & a_{22} \cdots & a_{2n} & 0 & 0 \cdots & 0 \\ \cdot & & & & & \\ \cdot & & & & & \\ \cdot & & & & & \\ \cdot & & & & & \\ \cdot & & & & & \\ a_{n1} & a_{n2} \cdots & a_{nn} & 0 & 0 \cdots & 0 \\ -1 & 0 \cdots & 0 & b_{11} & b_{12} \cdots & b_{1n} \\ 0 & -1 \cdots & 0 & b_{21} & b_{22} \cdots & b_{2n} \\ 0 & 0 & 0 & b_{31} & b_{32} \cdots & b_{3n} \\ \cdot & & \cdot & \cdot & \cdot & \cdot \\ \cdot & & \cdot & \cdot & \cdot & \cdot \\ \cdot & & \cdot & \cdot & \cdot & \cdot \\ \cdot & & \cdot & \cdot & \cdot & \cdot \\ 0 & 0 \ldots & -1 & b_{n1} & b_{n2} \cdots & b_{nn} \end{vmatrix} = |D|, \text{ say.}$$

In this case we have taken the C of theorem 4.5.2 to be

$$\begin{pmatrix} -1 & 0 & \cdots & 0 \\ 0 & -1 & \cdots & 0 \\ 0 & 0 & -1 \ldots & 0 \\ \cdot & \cdot & \cdot & \\ \cdot & \cdot & \cdot & \\ \cdot & \cdot & \cdot & \\ 0 & 0 & 0 \ldots & -1 \end{pmatrix}.$$

To the $(n+1)$th column of $|D|$ add b_{11} x first column, b_{21} x second column, ..., b_{n1} x nth column. After these operations we have:

$$|D| = |A||B| = \begin{vmatrix} a_{11} & a_{12} \cdots a_{1n} & c_{11} & 0 \cdots 0 \\ a_{21} & a_{22} \cdots a_{2n} & c_{21} & 0 \cdots 0 \\ \vdots & \vdots & \vdots & \vdots \\ a_{n1} & a_{n2} \cdots a_{nn} & c_{n1} & 0 \cdots 0 \\ -1 & 0 \cdots 0 & 0 & b_{12} \cdots b_{1n} \\ 0 & -1 \cdots 0 & 0 & b_{22} \cdots b_{2n} \\ \vdots & & \vdots & \vdots \\ 0 & 0 \cdots -1 & 0 & b_{n2} \cdots b_{nn} \end{vmatrix}$$

Now to the $(n+2)$th column add b_{12} x first column, b_{22} x second column, ..., b_{n2} x nth column. At this stage we have:

$$|A||B| = \begin{vmatrix} a_{11} & a_{12} \cdots a_{1n} & c_{11} & c_{12} & 0 \cdots 0 \\ \vdots & \vdots & \vdots & \vdots & \vdots \\ a_{n1} & a_{n2} \cdots a_{nn} & c_{n1} & c_{n2} & 0 \cdots 0 \\ -1 & 0 \cdots 0 & 0 & 0 & b_{13} \cdots b_{1n} \\ \vdots & \vdots & \vdots & \vdots & \vdots \\ 0 & 0 \cdots -1 & 0 & 0 & b_{n3} \cdots b_{nn} \end{vmatrix}$$

We continue this procedure until we arrive eventually at:

$$|A||B| = \begin{vmatrix} A & C \\ \begin{matrix} -1 & 0 \cdots 0 \\ 0 & -1 \cdots 0 \\ \vdots & \vdots \\ 0 \cdots & -1 \end{matrix} & O \end{vmatrix}$$

, where O is the $n \times n$ zero matrix.

$$\text{Thus} \quad |A||B| = \begin{vmatrix} A & C \\ -I_n & \bigcirc \end{vmatrix} = (-1)^{n^2} \begin{vmatrix} -I_n & \bigcirc \\ A & C \end{vmatrix}$$

by interchanging rows suitably.

$$\text{Hence} \quad |A||B| = (-1)^{n^2} \ |-I_n||C|$$

by theorem 4.5.2.

$$\text{Now } (-1)^{n^2} |-I_n||C| = (-1)^{n^2+n} |C|.$$

But $n(n + 1)$ is even, hence

$$(-1)^{n(n+1)} |C| = |C|.$$

$$\text{Thus finally} \quad |A||B| = |C| = |AB|$$

as required.

4.6 THE INVERSE OF A MATRIX

Let $A = (a_{ij})$ be an $n \times n$ matrix over the field F. As usual let A_{ij} be the cofactor of a_{ij} in A. Then we define the **adjugate** or **adjoint matrix** of A to be the matrix

$$\text{adj } A = (A_{ij})' = \begin{pmatrix} A_{11} & A_{21} \ldots\ldots & A_{n1} \\ A_{12} & A_{22} \ldots\ldots & A_{n2} \\ \cdot & \cdot & \cdot \\ \cdot & \cdot & \cdot \\ \cdot & \cdot & \cdot \\ A_{1n} & A_{2n} \ldots\ldots & A_{nn} \end{pmatrix}$$

$$\text{We have } A \,.\text{adj } A = \begin{pmatrix} a_{11} \ldots\ldots & a_{1n} \\ \cdot & \\ \cdot & \\ a_{n1} \ldots\ldots & a_{nn} \end{pmatrix} \begin{pmatrix} A_{11} \ldots\ldots & A_{n1} \\ \cdot & \\ \cdot & \\ A_{1n} \ldots\ldots & A_{nn} \end{pmatrix}$$

$$= \begin{pmatrix} |A| & 0 \ldots\ldots & 0 \\ 0 & |A| \ldots\ldots & 0 \\ \cdot & \cdot \ \cdots\cdots & \\ 0 & 0 \ \ldots\ldots & |A| \end{pmatrix} = |A| \begin{pmatrix} 1 & 0 \ldots\ldots & 0 \\ 0 & 1 \ldots\ldots & 0 \\ \cdot & \cdot \ \cdots\cdots & \cdot \\ 0 & 0 \ldots\ldots & 1 \end{pmatrix}$$

$$= |A| \ I_n.$$

Thus if $|A| \neq 0$ we have $AB = I_n$, where $B = (1/|A|)\,\text{adj}\,A$.

Similarly $BA = I_n$.

Note that in the above we have used the fact that

$$a_{i1}A_{j1} + a_{i2}A_{j2} + \ldots\ldots + a_{in}A_{jn} = 0 \text{ if } i \neq j,$$
$$= |A| \text{ if } i = j.$$

The latter is of course theorem 4.5.1. The former follows from the observation that the left hand side is the expansion of a determinant in which rows i and j are the same. If this cannot be seen at once the reader is advised to write the determinant out in some detail. Similar results hold for columns. Thus

$$a_{1i}A_{1j} + a_{2i}A_{2j} + \ldots\ldots + a_{ni}A_{nj} = 0 \text{ if } i \neq j,$$
$$= |A| \text{ if } i = j.$$

Now suppose that $AC = I_n$. Then $BAC = B$, with $BA = I_n$. Thus $I_n C = B$. Hence $C = B$.

From this we deduce that the matrix B satisfying $AB = BA = I_n$ is **unique**. To sum up, if $|A| \neq 0$ we have a unique matrix B satisfying $BA = AB = I_n$.

We call B the inverse of A and write $B = A^{-1}$.

Note that if A has an inverse then any matrix C satisfying $AC = I_n$ or $CA = I_n$ must be A^{-1}. Just multiply either equation by A^{-1} on the appropriate side.

Moreover if A has an inverse then $|A|$ must be non-zero. This follows from $I_n = AA^{-1}$. We have, taking determinants of both sides of this equation, $1 = |AA^{-1}| = |A||A^{-1}| \neq 0$. Thus $|A| \neq 0$.

Altogether we have proved the following

THEOREM 4.6.1 Let A be an n x n matrix over the field F. Then A has an inverse A^{-1} if and only if $|A| \neq 0$. If the inverse does exist then any matrix C satisfying $AC = I_n$ or $CA = I_n$ is the inverse A^{-1}. One expression for A^{-1} is $(1/|A|)\,\text{adj}\,A$. It is usual to call a matrix A **singular** if $|A| = 0$ and **non-singular** if $|A| \neq 0$.

Let S denote the set of all n x n matrices over the field F which have inverses. Suppose A and B are any two such matrices. Then $(AB)(B^{-1}A^{-1}) = A(BB^{-1})A^{-1} = A\,I_n\,A^{-1} = AA^{-1} = I_n$. Thus AB has an inverse, namely $B^{-1}A^{-1}$. Hence $AB \in S$. This means that matrix multiplication is a valid product in the set S. We note that $(A^{-1})^{-1} = A$. Hence if $A \in S$ then $A^{-1} \in S$. Thus each matrix in the set S has an inverse **in** S. The associative rule is obeyed by matrix multiplication as we know already. Altogether we have shown that S is a group under ordinary matrix multiplication. This group is called the **general linear group** and is denoted by $GL_n(F)$. It is of very great importance both in pure mathematics and in applications to physics and chemistry.

4.7 SOLUTION OF EQUATIONS

Let $A\boldsymbol{x} = \boldsymbol{k}$ be a system of n simultaneous equations in n variables over the field F. Thus A is an $n \times n$ matrix over F and $\boldsymbol{k}$ is an $n \times 1$ matrix or column vector also over F. Written out in more detail we have:

$$\begin{array}{ccccccc} a_{11}\, x_1 & + & a_{12}\, x_2 & + \ldots + & a_{1n}\, x_n & = & k_1 \\ a_{21}\, x_1 & + & a_{22}\, x_2 & + \ldots + & a_{2n}\, x_n & = & k_2 \\ \cdot & \cdot & \cdot & \cdot & \cdot\ \cdot & & \cdot \\ \cdot & \cdot & \cdot & \cdot & \cdot\ \cdot & & \cdot \\ \cdot & \cdot & \cdot & \cdot & \cdot\ \cdot & & \cdot \\ a_{n1}\, x_1 & + & a_{n2}\, x_2 & + \ldots + & a_{nn}\, x_n & = & k_n. \end{array}$$

If $|A| \neq 0$ then we can find A^{-1}. Now multiply the equation $A\boldsymbol{x} = \boldsymbol{k}$ by A^{-1} on the left. We get:

$$A^{-1}\, A\boldsymbol{x} = A^{-1}\, \boldsymbol{k}.$$

Thus $\qquad I_n\boldsymbol{x} = A^{-1}\, \boldsymbol{k}.$

Finally $\qquad \boldsymbol{x} = A^{-1}\, \boldsymbol{k}.$

A second method expresses the solution in terms of determinants. Consider again the system of n equations written out in some detail.

$$\begin{array}{ccccccccc} a_{11}\, x_1 & + & a_{12}\, x_2 & + \ldots + & a_{1i}\, x_i & + \ldots + & a_{1n}\, x_n & = & k_1 \\ a_{21}\, x_1 & + & a_{22}\, x_2 & + \ldots + & a_{2i}\, x_i & + \ldots + & a_{2n}\, x_n & = & k_2 \\ \cdot & \cdot & \cdot & \cdot & \cdot\ \cdot & \cdot & \cdot\ \cdot & & \cdot \\ \cdot & \cdot & \cdot & \cdot & \cdot\ \cdot & \cdot & \cdot\ \cdot & & \cdot \\ \cdot & \cdot & \cdot & \cdot & \cdot\ \cdot & \cdot & \cdot\ \cdot & & \cdot \\ a_{j1}\, x_1 & + & a_{j2}\, x_2 & + \ldots + & a_{ji}\, x_i & + \ldots + & a_{jn}\, x_n & = & k_j \\ \cdot & \cdot & \cdot & \cdot & \cdot\ \cdot & \cdot & \cdot\ \cdot & & \cdot \\ \cdot & \cdot & \cdot & \cdot & \cdot\ \cdot & \cdot & \cdot\ \cdot & & \cdot \\ \cdot & \cdot & \cdot & \cdot & \cdot\ \cdot & \cdot & \cdot\ \cdot & & \cdot \\ \cdot & \cdot & \cdot & \cdot & \cdot\ \cdot & \cdot & \cdot\ \cdot & & \cdot \\ a_{n1}\, x_1 & + & a_{n2}\, x_2 & + \ldots + & a_{ni}\, x_i & + \ldots + & a_{nn}\, x_n & = & k_n. \end{array}$$

Now multiply the jth equation by A_{ji}, the cofactor of a_{ji} in A, and **add** the resulting n equations. We get:

$$(\sum_{j=1}^{n} a_{j1} A_{ji}) x_1 + (\sum_{j=1}^{n} a_{j2} A_{ji}) x_2 + \ldots\ldots + (\sum_{j=1}^{n} a_{ji} A_{ji}) x_i + \ldots\ldots +$$

$$(\sum_{j=1}^{n} a_{jn} A_{ji}) x_n = \sum_{j=1}^{n} k_j A_{ji}.$$

Each of $\quad \sum_{j=1}^{n} a_{jr} A_{ji}$

is zero except for $\quad \sum_{j=1}^{n} a_{ji} A_{ji}$

which is $\quad |A|$.

Let $A^{(i)}$ be the matrix obtained from A by replacing the ith column by

$$k = \begin{pmatrix} k_1 \\ k_2 \\ \cdot \\ \cdot \\ \cdot \\ k_n \end{pmatrix}.$$

Then $\sum_{j=1}^{n} k_j A_{ji}$ is precisely $|A^{(i)}|$.

Thus the above sum of n equations reduces to:

$$|A|\, x_i = |A^{(i)}|.$$

If $\quad |A| \neq 0$, then we have: $x_i = |A^{(i)}|/|A|$.

In performing actual calculations the first method is more efficient than the second, particularly if A stays fixed and only k alters. Shortly we shall be looking at an even more efficient method. However the second method above does have some theoretical importance.

EXAMPLE 4.7.1 We solve the system of equations

$$\begin{aligned} x_1 + 2x_2 + x_3 &= 2 \\ x_1 - x_2 - x_3 &= 1 \\ 2x_1 + x_2 - x_3 &= 4 \end{aligned}$$

over the field **Q** of rationals.

Here $A = \begin{pmatrix} 1 & 2 & 1 \\ 1 & -1 & -1 \\ 2 & 1 & -1 \end{pmatrix}$ and $\mathbf{k} = \begin{pmatrix} 2 \\ 1 \\ 4 \end{pmatrix}$.

First Method.

Matrix of cofactors is: $\begin{pmatrix} 2 & -1 & 3 \\ 3 & -3 & 3 \\ -1 & 2 & -3 \end{pmatrix}$

$$|A| = 1 \cdot 2 + 2(-1) + 1 \cdot 3 = 2 - 2 + 3 = 3.$$

Adjugate of $A = \begin{pmatrix} 2 & 3 & -1 \\ -1 & -3 & 2 \\ 3 & 3 & -3 \end{pmatrix}$.

Thus $A^{-1} = (1/|A|) \operatorname{adj} A = \frac{1}{3}\begin{pmatrix} 2 & 3 & -1 \\ -1 & -3 & 2 \\ 3 & 3 & -3 \end{pmatrix}$

Solution is: $A^{-1}\mathbf{k} = \frac{1}{3}\begin{pmatrix} 2 & 3 & -1 \\ -1 & -3 & 2 \\ 3 & 3 & -3 \end{pmatrix}\begin{pmatrix} 2 \\ 1 \\ 4 \end{pmatrix} = \begin{pmatrix} 1 \\ 1 \\ -1 \end{pmatrix}$.

In terms of individual values: $x_1 = 1$, $x_2 = 1$, $x_3 = -1$.

Second Method.

We may write immediately:

$$x_1 = |A^{(1)}|/|A|, \qquad x_2 = |A^{(2)}|/|A|, \qquad x_3 = |A^{(3)}|/|A|.$$

As above $|A| = 3$. We expand $|A^{(1)}|$ by the first column, $|A^{(2)}|$ by the second column and $|A^{(3)}|$ by the third column using the appropriate cofactors given in the matrix of cofactors above. We obtain:

$$|A^{(1)}| = \begin{vmatrix} 2 & 3 & -1 \\ 1 & -3 & 2 \\ 4 & 3 & -3 \end{vmatrix} = 2(2) + 1(3) + 4(-1) = 4 + 3 - 4 = 3.$$

$$|A^{(2)}| = \begin{vmatrix} 1 & 2 & 1 \\ 1 & 1 & -1 \\ 2 & 4 & -1 \end{vmatrix} = 2(-1) + 1(-3) + 4(2) = -2 - 3 + 8 = 3.$$

$$|A^{(3)}| = \begin{vmatrix} 1 & 2 & 2 \\ 1 & -1 & 1 \\ 2 & 1 & 4 \end{vmatrix} = 2(3) + 1(3) + 4(-3) = 6 + 3 - 12 = -3.$$

Thus $x_1 = 3/3 = 1, \quad x_2 = 3/3 = 1, \quad x_3 = -3/3 = -1.$

4.8 VECTORS AND VECTOR SPACES

It is assumed that the reader has met vectors in 2 and 3-dimensions such as are described in 'Analytic Geometry and Vectors' by J. Hunter (Blackie and Chambers 1972). Here we recall the concept of a vector in 3-dimensions. Throughout this book the term Cartesian coordinates will refer to a system of straight line axes in space which are at right angles to each other. We denote by $\boldsymbol{i}, \boldsymbol{j}, \boldsymbol{k}$ vectors from the origin, of unit length, directed along the x, y, z or x_1, x_2, x_3, axes respectively. Any vector $\boldsymbol{v}$ from the origin may be written in the form:

$$\boldsymbol{v} = x\boldsymbol{i} + y\boldsymbol{j} + z\boldsymbol{k}.$$

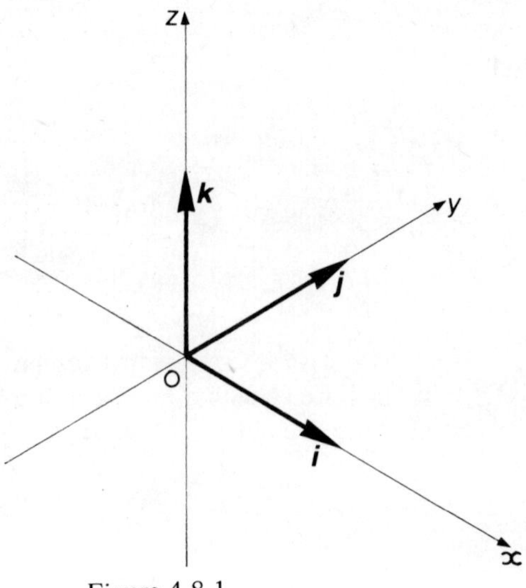

Figure 4.8.1

Addition.

If $\quad u = x'i + y'j + z'k$

then the sum of u and v is defined to be:

$$u + v = (x' + x)\,i + (y' + y)\,j + (z' + z)\,k$$

Multiplication by a Scalar. If a is a real number we define the product av to be $(ax)\,i + (ay)\,j + (az)\,k$.

Another example of a similar system is the set of all $n \times 1$ matrices or **column** vectors, with addition, and multiplication by a scalar, as defined already for matrices. Similar remarks apply to the set of all $1 \times n$ matrices or **row** vectors over some field F.

Yet another example of the kind of system we have in mind is provided by the set of all real functions f defined on some interval, say $[a, b]$, where $a, b \in \mathbf{R}$.

Thus $\quad f : [a, b] \longrightarrow \mathbf{R}.$

If f and g are two such functions we define their sum to be the function $f + g$ with the property that $(f+g)(x) = f(x) + g(x)$ for all $x \in [a, b]$.

If $a \in \mathbf{R}$ we define multiplication of f by the scalar a to be the function (af) with the property that $(af)(x) = a(f(x))$.

All these systems behave in a similar way with respect to addition and multiplication by a scalar as given above. It would be a great duplication of effort if we proved the properties of each of these systems separately. For this reason we define an abstract system which applies to each of our examples as a special case. In fact this does more than avoid duplication of effort. It enables us to see more clearly the properties of our systems by leaving out the irrelevant facts that serve only to make the situation more complex. The abstract system that serves our purpose is called a **vector space over a field**. We turn to its formal definition.

DEFINITION 4.8.1 Let V be a set whose elements are called **vectors**. Let F be a field. (In subsequent work this will usually be either **R** or **C**).

A binary operation of addition is defined on V relative to which V is an abelian group. (See Definition 4.3.1). A function $f : F \times V \longrightarrow V$ defines multiplication of a vector by a scalar. We denote $f(a, v)$ by av, and call this vector the product of v by the scalar a. This product must satisfy the following axioms:

(1) $a(v_1 + v_2) = av_1 + av_2$, $a \in F$; v_1 , $v_2 \in V$.

(2) $(a + b)v = av + bv$, $a, b \in F$; $v \in V$.

(3) $(ab)v = a(bv)$, $a, b \in F$; $v \in V$.

(4) $1v = v$, $1 \in F$; $v \in V$.

In (4) the 1 denotes the multiplicative identity of the field F. Of course if the field is **Q**, **R**, or **C** this is just the ordinary number 1.

The reader should have no difficulty in checking that all the systems mentioned earlier do indeed satisfy these axioms and therefore are particular examples of a vector space.

From the axioms we deduce some expected properties. We collect these in:

THEOREM 4.8.1 Let 0 denote the zero of the field F. Let $\mathit{0}$ be the **additive identity** (the zero) of the additive group V. The inverse of v in the group V is denoted by $-v$. Then we have:

(1) $0v = \mathit{0}, v \in V$.

(2) $a\mathit{0} = \mathit{0}, a \in F$.

(3) $-(-v) = v, v \in V$.

(4) $(-a)v = -(av)$.

(5) $a(-v) = -(av)$.

(6) $(-a)(-v) = av$.

PROOF

(1) $(0 + 0)\,v \quad = \quad 0v + 0v$

$\quad\;\; 0\,v \quad = \quad 0v + 0v$

$$
\begin{aligned}
0v + (-(0v)) &= (0v + 0v) + (-(0v)) \\
0 &= 0v + (0v + (-(0v))) \\
0 &= 0v + 0 = 0v.
\end{aligned}
$$

(2)
$$
\begin{aligned}
a(0 + 0) &= a0 \\
a0 + a0 &= a0 \\
(a0 + a0) + (-(a0)) &= a0 + (-(a0)) \\
a0 + (a0 + (-(a0))) &= 0 \\
a0 + 0 &= 0 \\
a0 &= 0.
\end{aligned}
$$

(3)
$$
\begin{aligned}
(-v) + v &= 0 \\
(-(-v)) + ((-v) + v) &= -(-v) \\
(-(-v) + (-v)) + v &= -(-v) \\
0 + v &= -(-v) \\
v &= -(-v).
\end{aligned}
$$

(4)
$$
\begin{aligned}
(a + (-a))\, v &= av + (-a)\, v \\
0v &= av + (-a)\, v \\
0 &= av + (-a)\, v \\
-(av) &= -(av) + (av + (-a)\, v) \\
-(av) &= (-(av) + av) + (-a)\, v \\
-(av) &= 0 + (-a)\, v \\
-(av) &= (-a)\, v.
\end{aligned}
$$

(5)
$$
\begin{aligned}
a(v + (-v)) &= av + a(-v) \\
a0 &= av + a(-v) \\
0 &= av + a(-v) \\
-(av) &= -(av) + (av + a(-v)) \\
-(av) &= (-(av) + av) + a(-v) \\
-(av) &= 0 + a(-v) \\
-(av) &= a(-v).
\end{aligned}
$$

(6)
$$
\begin{aligned}
(-a)(-v) &= -(a(-v)) = -(-(av)) \\
&= av,
\end{aligned}
$$

where use has been made of (4), (5), and (3).

We now require several definitions before we can prove some important results.

DEFINITION 4.8.2 Let V be a vector space over a field F. The vectors $v_1, v_2, \ldots \ldots, v_n$ are called **generators** of V, or are said to **generate** V, if every $v \in V$ can be written $v = a_1 v_1 + a_2 v_2 + \ldots\ldots + a_n v_n$ for some $a_1, a_2, a_3, \ldots\ldots, a_n \in F$. We say that V is **finitely generated** if such a finite set $\{v_1, v_2, \ldots, v_n\}$ exists. We write $< v_1, v_2, v_3, \ldots\ldots, v_n >$ for the space generated by the v_i, $i = 1, \ldots\ldots, n$.

DEFINITION 4.8.3 Let V be a vector space over a field F. The vectors $v_1, v_2, \ldots\ldots, v_r$ are said to be **linearly dependent** if $a_1 v_1 + a_2 v_2 + \ldots\ldots + a_r v_r = 0$ for some $a_i \in F$ where **not all** the a_i are zero.

If **no** such expression exists unless **all** the a_i are zero, then we say that the vectors $v_1, v_2, \ldots\ldots, v_r$ are **linearly independent**.

We say that v is **linearly dependent on** the vectors

$$v_1, v_2, \ldots\ldots, v_r \text{ if } v = a_1 v_1 + a_2 v_2 + \ldots\ldots + a_r v_r$$

for some $a_i \in F$.

DEFINITION 4.8.4 Any linearly independent generating set of V is called a **BASIS** of V.

Note that at least one basis of V exists if V is finitely generated. Just take a generating set $\{v_1, v_2, \ldots\ldots, v_m\}$ and discard the unwanted generators. If V is finitely generated, and so has a basis with a finite number of vectors, we say that V is of **finite dimension**.

We turn now to a sequence of theorems of fundamental importance in the theory of vector spaces.

THEOREM 4.8.2 The non-zero vectors $v_1, v_2, \ldots\ldots, v_n$ are linearly dependent if and only if some v_k, $2 \leqq k \leqq n$, is a linear combination of the preceding ones.

By **linear combination** of vectors $w_1, w_2, \ldots\ldots, w_m$ we mean any expression of the form $a_1 w_1 + a_2 w_2 + \ldots\ldots + a_m w_m$ where the a_i are scalars.

PROOF Let $v_1, v_2, \ldots\ldots, v_n$ be linearly dependent. Let k be the first integer $2 \leqq k \leqq n$ for which $v_1, v_2, \ldots\ldots, v_k$ are linearly dependent.

Then $a_1 v_1 + a_2 v_2 + \ldots\ldots + a_k v_k = 0$ with not all the a_i zero. Moreover by definition of k we have $a_k \neq 0$.

Thus $\quad v_k = (-a_1/a_k) v_1 + (-a_2/a_k) v_2 + \ldots\ldots + (-a_{k-1}/a_k) v_{k-1}.$

Conversely if $v_k = a_1 v_1 + \ldots\ldots + a_{k-1} v_{k-1}$, then

$$0 = a_1 v_1 + \ldots\ldots + a_{k-1} v_{k-1} + (-1) v_k + \ldots + a_n v_n.$$

Hence, since (-1) at least is not zero, the vectors $v_1, v_2, \ldots\ldots, v_n$ are linearly dependent.

THEOREM 4.8.3 Let V be a finite dimensional vector space over the field F. Let $v_1, v_2, \ldots\ldots, v_m$ be any set of linearly independent vectors in V, not already a basis. Then we can find vectors $v_{m+1}, \ldots\ldots, v_{m+p}$ from V such that $\{v_1, v_2, \ldots\ldots, v_m, v_{m+1}, \ldots\ldots, v_{m+p}\}$ is a basis.

PROOF Since V is finite dimensional, it has a finite basis, say $\{w_1, w_2, \ldots\ldots, w_n\}$.

Consider the set $S = \{v_1, v_2, \ldots\ldots, v_m, w_1, \ldots\ldots, w_n\}$. The set S is linearly dependent because all the v_i are linearly dependent on the w_j. Thus some vector of S is dependent on the preceding ones by the previous theorem. Let z be the first such vector. Now $z \neq v_i$ for any i, because the v_i are linearly independent. Thus $z = w_i$ for some i.

Now consider $\{v_1, v_2, \ldots\ldots, v_m, w_1, \ldots\ldots, w_{i-1}, w_{i+1}, \ldots\ldots, w_n\}$. Every vector of V can be expressed as a linear combination of these vectors. If these vectors are linearly independent, we have our basis. If not, we repeat the previous argument.

Eventually we arrive at a basis containing $\{v_1, v_2, \ldots\ldots, v_m\}$.

THEOREM 4.8.4 The number of elements in any basis of a finite dimensional vector space V is the same.

PROOF Let $\{v_1, v_2, \ldots\ldots, v_m\}$ and $\{w_1, w_2, \ldots\ldots, w_n\}$ be any two bases of V. Suppose that $m > n$.

Write $\{v_m, w_1, w_2, \ldots\ldots, w_n\}$. Then this is a set of linearly dependent vectors, since v_m is dependent on the basis vectors $w_1, \ldots\ldots, w_n$.

Now use the argument of the previous theorem to show that $\{v_m, w_1, \ldots\ldots, w_{i-1}, w_{i+1}, \ldots\ldots, w_n\}$ (w_i omitted) generates V. It follows that v_{m-1} is a linear combination of $v_m, w_1, \ldots\ldots, w_{i-1}, w_{i+1}, \ldots\ldots, w_n$. Hence the set $\{v_{m-1}, v_m, w_1, \ldots\ldots, w_{i-1}, w_{i+1}, \ldots\ldots, w_n\}$ is linearly dependent. We repeat the argument. At each stage we discard a redundant w_j.

Eventually $\{v_i, v_{i+1}, v_{i+2}, \ldots\ldots, v_m\}$ will be linearly dependent for some i. This contradicts the hypothesis that $\{v_1, v_2, \ldots\ldots, v_m\}$ is a basis of V.

We conclude that $m \not> n$.

Similarly, interchanging the roles of the two bases in the above, we deduce that $n \not> m$.

Altogether we have $m = n$.

The above result enables us to give the following definition.

DEFINITION 4.8.5 Let V be a finitely generated vector space over the field F. The number of vectors in any basis of V is called the DIMENSION of V and is denoted by dim V.

THEOREM 4.8.5 Let V be a vector space of dimension n over the field F. Then

(1) every set of $n + 1$ vectors is linearly dependent,

(2) a set of n vectors of V is a basis of V if and only if the vectors are linearly independent,

(3) a set of n vectors of V is a basis of V if and only if the set generates V.

PROOF

(1) If the $n + 1$ vectors are linearly independent then they can be included in a basis by theorem 4.8.3. This basis has at least $n + 1$ vectors. But dim $V = n$. This contradiction shows that any $n + 1$ vectors in V must be linearly dependent.

(2) If any n vectors are linearly independent they can be included in a basis (by theorem 4.8.3) which has n members because dim $V = n$. Thus the original n vectors form a basis. The converse is immediate.

(3) If n vectors generate V then they include a basis; just discard unwanted vectors. But this basis must have n vectors in it. Thus the original set of n vectors must form a basis. Again the converse is trivial.

4.9 SUBSPACES

In geometry a line is a subspace of a plane which in turn is a subspace of the whole 3-dimensional space. We formalise these intuitive ideas by the following:

DEFINITION 4.9.1 A non-empty subset S of a vector space V is called a **subspace** of V if S is itself a vector space over the same field with the **same** addition and multiplication by a scalar as V.

Note that for each of the abstract structures mentioned already, i.e. ring, field, and group, there is the corresponding concept of subring, subfield and subgroup. The definition in each case parallels that for a subspace of a vector space.

The following gives a simple test for a subset to be a subspace.

THEOREM 4.9.1 Let V be a vector space over the field F. Let S be a non-empty subset of V. Then S is a subspace of V if and only if

(1) $v_1, v_2 \in S \longrightarrow v_1 + v_2 \in S$,

(2) $a \in F$ and $v \in S \longrightarrow av \in S$.

PROOF

If S is a subspace of V then the addition in V is a valid addition for S and the multiplication by a scalar is a valid operation for S. This leads at once to (1) and (2).

Conversely suppose (1) and (2) hold. Then these conditions show that the addition in V is a valid addition for S and similarly for the multiplication by a scalar. Now take any $v \in S$. By (2) $(-1)v = -v$ is in S. Thus any $v \in S$ has an inverse in S. But then $v + (-v) \in S$ by (1). Hence $0 \in S$. Thus S has an additive identity. All the other axioms for a vector space are satisfied by the vectors of S because these vectors are also vectors of V. We describe this situation by saying that S inherits the axioms from V. Thus S is a subspace of V.

Seemingly obvious is the following:

THEOREM 4.9.2 Let V be a vector space of dimension n over the field F. Let W be any subspace of V. Then $\dim W \leqq n$. We have $\dim W = n \leftrightarrow W = V$.

PROOF

If $W = O$, the zero subspace, then $\dim W = 0$. If W contains a non-zero vector v_1, let $W = \langle v_1 \rangle$. Then $\dim W_1 = 1$ and $W_1 \subset W$. If $W_1 \neq W$, let $v_2 \in W$ with $v_2 \notin W_1$. Then $W_2 = \langle v_1, v_2 \rangle$ is 2-dimensional. By theorem 4.8.5 (1) we cannot find $n + 1$ linearly independent vectors. Thus after at most n steps, the above procedure must terminate. Thus $\dim W \leqq n$.

If $\dim W = n$ then a basis for W is a basis for V.

The following theorem shows that any basis of a subspace of V can be extended to a basis of the whole space.

THEOREM 4.9.3 Let V be a vector space of dimension n over the field F. Let W be a subspace of V of dimension m. Let $\{w_1, w_2, \ldots\ldots, w_m\}$ be any basis of W. Then there exist vectors $w_{m+1}, w_{m+2}, \ldots\ldots, w_n$ in V such that $\{w_1, w_2, w_3, \ldots\ldots, w_n\}$ is a basis of V.

PROOF

The required result follows at once from theorem 4.8.3.

DEFINITION 4.9.2 As noted in section 4.8. the set of all $1 \times n$ matrices, or row vectors, over a field F is a vector space under the addition of matrices and multiplication by a scalar. We call this vector space the space of n-tples over F and denote it by F^n. It turns out to be the model for all vector spaces of dimension n over F. (Similar remarks apply for $n \times 1$ matrices, or column vectors).

Before we can state this precisely we need to define the concept of isomorphism of vector spaces.

We have already mentioned this idea in connection with rings; see definition 3.6.1. As with rings isomorphic vector spaces are indistinguishable as regards the operations of + and multiplication by a scalar. The only difference is the name given to the individual vectors. Formally we have:

DEFINITION 4.9.3 Let f be a function from the vector space V over the field F

to the vector space W over the same field F. $f : V \longrightarrow W$. If

(1) $f(v_1 + v_2) = f(v_1) + f(v_2)$

(2) $f(av) \quad = af(v)$

for all $a \in F$ and $v_1, v_2, v \in V$; then f is said to be a **homomorphism** from V to W. Usually the term **linear transformation** is used rather than homomorphism. From now on we shall follow this convention. If f is **one-one** and **onto** (a bijection) then f is said to be an **isomorphism** from V to W, and we say that V and W are isomorphic vector spaces, written $V \cong W$.

We are now in a position to state and prove

THEOREM 4.9.4 Let V be a vector space of dimension n over the field F. Let F^n be the vector space of n-tples over F. Then V and F^n are isomorphic. $V \cong F^n$.

PROOF

Let $\{v_1, v_2, \ldots\ldots, v_n\}$ be a basis of V.

Define a function f: $F^n \longrightarrow V$ by

$$f((a_1, a_2, \ldots\ldots, a_n)) = a_1 v_1 + a_2 v_2 + \ldots\ldots + a_n v_n,$$

where $a_1, a_2, \ldots\ldots, a_n$ are from F.

Since the set $\{v_1, v_2, \ldots\ldots, v_n\}$ generates V the function f is onto.

Now suppose that

$$f((a_1, a_2, \ldots, a_n)) = f((b_1, b_2, \ldots, b_n)).$$

Then $\quad a_1 v_1 + a_2 v_2 + a_3 v_3 + \ldots\ldots + a_n v_n = b_1 v_1 + b_2 v_2 + \ldots + b_n v_n$.

Thus $\quad (a_1 v_1 + \ldots\ldots + a_n v_n) - (b_1 v_1 + \ldots + b_n v_n) = 0$.

Hence $\quad (a_1 v_1 - b_1 v_1) + (a_2 v_2 - b_2 v_2) + \ldots\ldots + (a_n v_n - b_n v_n) = 0$.

Finally $\quad (a_1 - b_1) v_1 + (a_2 - b_2) v_2 + \ldots\ldots + (a_n - b_n) v_n = 0$.

However the vectors $v_1, v_2, \ldots\ldots, v_n$ are linearly independent. Thus the coefficients $a_i - b_i$ must all be zero.

Hence $\quad a_i = b_i$ for all i.

It follows that f is one-one.

We have:

$$f((a_1, a_2, \ldots\ldots, a_n) + (b_1, b_2, \ldots\ldots, b_n)) =$$

$$= f(((a_1 + b_1), \ldots\ldots, (a_n + b_n))) =$$

$$= (a_1 + b_1)v_1 + (a_2 + b_2)v_2 + \ldots\ldots + (a_n + b_n)v_n =$$

$$= a_1 v_1 + a_2 v_2 + \ldots\ldots + a_n v_n + b_1 v_1 + \ldots\ldots + b_n v_n =$$

$$= f((a_1, a_2, a_3, \ldots\ldots, a_n)) + f((b_1, b_2, \ldots\ldots, b_n))$$

and

$$f(a(a_1, a_2, a_3, \ldots\ldots, a_n)) = f((aa_1, aa_2, \ldots\ldots, aa_n))$$

$$= aa_1 v_1 + aa_2 v_2 + \ldots\ldots + aa_n v_n$$

$$= a(a_1 v_1 + a_2 v_2 + \ldots\ldots + a_n v_n)$$

$$= a f((a_1, a_2, \ldots\ldots, a_n)).$$

Thus f is a linear transformation.

Altogether f is an isomorphism.

Note that in the course of the above proof we proved that the expression of a vector v as a linear combination of basis vectors is **unique**. The converse is also true. If **any** vector v can be expressed as a linear combination of vectors $v_1, v_2, \ldots\ldots, v_n$ **in only one way** then the vectors $v_1, v_2, \ldots\ldots, v_n$ form a basis of V. Clearly the vectors $v_1, v_2, \ldots\ldots, v_n$ generate V. Now suppose that $a_1 v_1 + a_2 v_2 + \ldots\ldots + a_n v_n = 0$.

Then $\quad a_1 v_1 + a_2 v_2 + \ldots\ldots + a_n v_n = 0v_1 + 0v_2 + \ldots\ldots + 0v_n$.

By uniqueness $\quad a_1 = 0,\ a_2 = 0, \ldots\ldots, a_n = 0$.

Thus $\quad v_1, \ldots\ldots, v_n$ are linearly independent.

Hence $\{v_1, v_2, \ldots\ldots, v_n\}$ is a basis of V.

EXAMPLE 4.9.1

Let $V = \mathbf{R}^4$.

Let $W = \langle (1, 0, 1, 1), (-1, 2, 0, 1), (0, 1, 0, 1) \rangle$.

Find a basis of W and extend it to a basis of V.

Solution. First check that the vectors generating W form a basis of W. To do this, suppose that

$$a(1, 0, 1, 1) + b(-1, 2, 0, 1) + c(0, 1, 0, 1) = (0, 0, 0, 0)$$

where a, b, c are real numbers.

This leads to four equations for a, b, and c. The solution is $a = b = c = 0$. This proves that the given vectors form a basis of W.

Now $\{(1, 0, 0, 0), (0, 1, 0, 0), (0, 0, 1, 0), (0, 0, 0, 1)\}$

is a basis of V. Consider the set

$$\{(1, 0, 1, 1), (-1, 2, 0, 1), (0, 1, 0, 1), (1, 0, 0, 0), (0, 1, 0, 0), (0, 0, 1, 0), (0, 0, 0, 1)\}.$$

Now $(1, 0, 0, 0)$ is not linearly dependent on the given basis of W. Thus the set

$$\{(1, 0, 1, 1), (-1, 2, 0, 1), (0, 1, 0, 1), (1, 0, 0, 0)\}$$

is linearly independent. It contains four vectors.

By theorem 4.8.5 (2) this set is a basis for V. Since it includes the given basis of W, it is the required basis of V.

4.10 THE RANK OF A MATRIX

We now use our abstract ideas to study matrices more deeply.

$$\text{Let } A = \begin{pmatrix} a_{11} & a_{12} \dots & a_{1n} \\ \cdot & & \\ \cdot & & \\ a_{m1} & a_{m2} \dots & a_{mn} \end{pmatrix}$$

be an $m \times n$ matrix over the field F. We consider the rows of A as vectors in F^n and the columns of A as vectors in F^m.

$$\text{Write } \boldsymbol{a}_i = \begin{pmatrix} a_{1i} \\ \cdot \\ \cdot \\ \cdot \\ \cdot \\ a_{mi} \end{pmatrix} \quad \text{and } \boldsymbol{b}_j = (a_{j1}, \dots, a_{jn}).$$

DEFINITION 4.10.1 The vector space $\langle \boldsymbol{b}_1, \boldsymbol{b}_2, \ldots\ldots, \boldsymbol{b}_m \rangle$ is called the **row space** of the matrix A and the vector space $\langle \boldsymbol{a}_1, \boldsymbol{a}_2, \boldsymbol{a}_3, \ldots\ldots, \boldsymbol{a}_n \rangle$ is called the **column space** of A. The **dimension** of $\langle \boldsymbol{b}_1, \boldsymbol{b}_2, \ldots\ldots, \boldsymbol{b}_m \rangle$ is called the **row rank** of A and the **dimension** of $\langle \boldsymbol{a}_1, \boldsymbol{a}_2, \ldots\ldots, \boldsymbol{a}_n \rangle$ is called the **column rank** of A.

One immediately wonders if the two ranks are related. In fact the relation is simpler than might be expected.

THEOREM 4.10.1 Let A be an $m \times n$ matrix over the field F. Then the row rank of A = the column rank of A.

PROOF

Let the column rank of A be r. Let $\boldsymbol{a}_{k_1}, \boldsymbol{a}_{k_2}, \ldots\ldots, \boldsymbol{a}_{k_r}$ be r linearly independent columns.

Then $\quad \boldsymbol{a}_i = \sum_{h=1}^{r} p_{hi} \boldsymbol{a}_{k_h}$ for $i = 1,2,\ldots\ldots\ldots,n$.

Equating the jth coordinates of the vectors on each side of this equation we get

(1) $\quad a_{ji} = \sum_{h=1}^{r} p_{hi}\, a_{j,k_h}$ for $i = 1,2,3,\ldots\ldots,m$.

Let the rows of A be $\boldsymbol{b}_j$, $j = 1,2, \ldots\ldots, m$. Then a_{ji} is the ith element or coordinate of the row vector $\boldsymbol{b}_j$ and p_{hi} is the ith element of the row vector $(p_{h1}, p_{h2}, p_{h3}, \ldots\ldots, p_{hr}) = \boldsymbol{p}_h$, say.

Now rewrite the equation (1) above in the form:

$$\boldsymbol{b}_j = \sum_{h=1}^{r} \boldsymbol{p}_h\, a_{j,k_h} \text{ for } j = 1,2,3,\ldots\ldots,m.$$

Thus each row $\boldsymbol{b}_j$ of A can be written as a linear combination of the r vectors $\boldsymbol{p}_1, \boldsymbol{p}_2, \ldots\ldots, \boldsymbol{p}_r$.

Thus the row rank of $A \leqq r$ = column rank of A.

Apply the same argument to A'. This gives

(2) the row rank of $A' \leqq$ the column rank of A'.

However the row rank of A' = the column rank of A and the column rank of A' = the row rank of A.

Thus (2) above gives:

the column rank of $A \leqq$ the row rank of A.

Altogether we have:

the column rank of A = the row rank of A.

4.11 ELEMENTARY OPERATIONS ON VECTORS

We note three elementary operations on a generating set $S = \{v_1, v_2, \ldots\ldots, v_n\}$ of a vector space V.

(1) Replace $\{v_1, v_2, \ldots\ldots, v_n\}$ by $\{v_1, \ldots, v_{i-1}, v_i + v_j, v_{i+1}, \ldots, v_j, \ldots, v_n\}$

(2) Replace $\{v_1, v_2, \ldots\ldots, v_n\}$ by $\{v_1, v_2, \ldots, v_{i-1}, kv_i, v_{i+1}, \ldots, v_n\}$

where $0 \neq k \in F$ the field over which V is defined.

(3) Replace $\{v_1, v_2, \ldots\ldots, v_i, \ldots, v_j, \ldots\ldots, v_n\}$ by

$$\{v_1, \ldots, v_{i-1}, v_j, v_{i+1}, \ldots\ldots, v_{j-1}, v_i, v_{j+1}, \ldots, v_n\}.$$

Note that in each case, if S is the generating set originally and S' is the generating set after the elementary operation has been carried out, then $\langle S \rangle = \langle S' \rangle$. Thus the elementary operations do **not** alter the dimension of the space generated. In particular a set of linearly independent vectors remains linearly independent.

We apply these operations to the rows and columns of a matrix considered as vectors. In these cases they are called **elementary row** or **elementary column operations.**

These operations can be carried out by multiplying the matrix before or after (pre- or post- multiplying) by a matrix obtained by applying the elementary operation concerned to the appropriate identity matrix. These latter matrices are called elementary matrices. We note that elementary matrices are **non-singular**, that is they have inverses. This is the same as saying that they have non-zero determinants; see theorem 4.6.1. This follows from $|I| = 1 \neq 0$, and the elementary operations only alter the value of the determinant by $k \neq 0$ or ± 1, as a multiplying factor.

EXAMPLE 4.11.1

(1) Take $I = \begin{pmatrix} 1 & 0 & 0 \\ 0 & 1 & 0 \\ 0 & 0 & 1 \end{pmatrix}$ and add the second row

to the third row. We get the elementary matrix $E_1 = \begin{pmatrix} 1 & 0 & 0 \\ 0 & 1 & 0 \\ 0 & 1 & 1 \end{pmatrix}$.

Now consider the matrix $A = \begin{pmatrix} 2 & 3 & 4 \\ 5 & 6 & -1 \\ 0 & 1 & 0 \end{pmatrix}$

We have $E_1A = \begin{pmatrix} 2 & 3 & 4 \\ 5 & 6 & -1 \\ 5 & 7 & -1 \end{pmatrix}$

which is the effect of applying the elementary row operation: add the second row of A to the third row of A.

(2) Multiply the third column of I by 2 to get the elementary matrix

$$E_2 = \begin{pmatrix} 1 & 0 & 0 \\ 0 & 1 & 0 \\ 0 & 0 & 2 \end{pmatrix}$$

Now consider $AE_2 = \begin{pmatrix} 2 & 3 & 4 \\ 5 & 6 & -1 \\ 0 & 1 & 0 \end{pmatrix} \begin{pmatrix} 1 & 0 & 0 \\ 0 & 1 & 0 \\ 0 & 0 & 2 \end{pmatrix} = \begin{pmatrix} 2 & 3 & 8 \\ 5 & 6 & -2 \\ 0 & 1 & 0 \end{pmatrix}$

This is the result of multiplying the third column of A by 2.

Thus pre-multiplying by E_1 performs an elementary row operation and post-multiplying by E_2 performs an elementary column operation.

By using elementary operations any matrix may be transformed into a matrix of particularly simple form. For this we need the idea of the **partitioned** matrix. This is a matrix written as a matrix whose elements are themselves matrices.

Thus the matrix $A = \begin{pmatrix} 3 & 4 & 0 & -1 \\ 2 & 6 & 5 & 3 \\ 9 & 0 & 2 & -4 \end{pmatrix}$

may be written as the partitioned matrix:

$$A = \begin{pmatrix} A_{11} & A_{12} \\ A_{21} & A_{22} \end{pmatrix}, \text{ where } A_{11} = \begin{pmatrix} 3 & 4 \\ 2 & 6 \end{pmatrix},$$

$$A_{12} = \begin{pmatrix} 0 & -1 \\ 5 & 3 \end{pmatrix}, \quad A_{21} = (9 \quad 0),$$

$$A_{22} = (2 \quad -4).$$

Note that A_{11} and A_{21} have same number of columns, as do A_{12} and A_{22}, while A_{11} and A_{12} have the same number of rows as do A_{21} and A_{22}.

THEOREM 4.11.1 Let A be an $m \times n$ matrix of rank r. Then by elementary row and column operations A can be changed to a matrix of the form

$$\begin{pmatrix} I_r & O \\ O & O \end{pmatrix}$$

where I_r = the $r \times r$ identity matrix.

This means that non-singular matrices P and Q exist, namely the product of elementary matrices, such that

$$PAQ = \begin{pmatrix} I_r & O \\ O & O \end{pmatrix},$$

where the O's denote suitable zero matrices.

PROOF

$$\text{Let } A = \begin{pmatrix} a_{11} & a_{12} \cdots\cdots a_{1n} \\ a_{21} & a_{22} \cdots\cdots a_{2n} \\ \cdot & \\ \cdot & \\ \cdot & \\ a_{n1} & a_{n2} \cdots\cdots a_{nn} \end{pmatrix}.$$

Rearrange rows so that the row with a non-zero element x_i nearest to the first position in the row becomes the new first row. By taking suitable multiples of this row away from the other rows, we reduce all elements in the ith column to zero apart from x_i itself. Now we leave out the first row and repeat the procedure on the remaining $m - 1$ rows.

We continue to repeat this procedure until we get an echelon form:

$$\begin{pmatrix} 0 & 0\ldots\ldots X & X \ldots\ldots\ldots X & X & X & X \\ 0 & 0\ldots\ldots 0 & X \ldots\ldots\ldots X & X & X & X \\ 0 & 0\ldots\ldots 0 & 0 \ldots\ldots\ldots X & X & X & X \\ \cdot & \cdot \quad\quad \cdot & \cdot \quad\quad\quad \cdot & \cdot & \cdot & \cdot \\ 0 & 0\ldots\ldots 0 & 0 \ldots\ldots\ldots 0 & X & X & X \end{pmatrix}$$

where the crosses denote elements that may be non-zero.

We now use the first non-zero column to reduce all remaining elements in the first row to zero. Then we use the second non-zero column in a similar way.

We repeat the procedure and rearrange rows and columns in a suitable way until we have a matrix in which the only non-zero elements lie on the leading diagonal from the top left hand corner down. The matrix now looks like:

$$\begin{pmatrix} \times & 0 & 0 & 0 \ldots\ldots\ldots & 0 \\ 0 & \times & 0 & 0 \ldots\ldots\ldots & 0 \\ 0 & 0 & \times & 0 \ldots\ldots\ldots & 0 \\ \cdot & \cdot & \cdot & \cdot & \cdot \\ \cdot & \cdot & \cdot & \cdot & \cdot \end{pmatrix}$$

By multiplying by suitable scalars from the field F over which A is defined, we get the matrix:

$$\begin{pmatrix} I_t & O \\ O & O \end{pmatrix}.$$

Now the rank of $\begin{pmatrix} I_t & O \\ O & O \end{pmatrix}$ is t.

However all our operations consist of sequences of elementary operations which do not affect the rank of A. This follows from the definition of rank, theorem 4.10.1, and the remarks after the definition of elementary operations at the beginning of this section.

Thus $t = r$.

The following theorem gives information about the rank of non-singular matrices.

THEOREM 4.11.2 Let A be an $n \times n$ matrix over the field F. Then A is non-singular ($|A| \neq 0$) if and only if the rank of A is n.

PROOF

Let the rank of A be n. Then for some elementary matrices $E_1, E_2, E_3, \ldots\ldots$, E_m and $F_1, F_2, \ldots\ldots, F_t$ we have $I_n = E_1 \ldots E_m A F_{-1} \ldots F_t$, by theorem 4.11.1. Taking determinants of both sides of this equation we get

$$1 = |E_1|\,|E_2|\,|E_3| \ldots\ldots |A|\,|F_1|\,|F_2| \ldots\ldots |F_t|.$$

Now all the $|E_i|$ and $|F_j|$ are non-zero. Thus $|A| \neq 0$.
Conversely if $|A| \neq 0$, then suppose rank $A < n$.
We have $E_1 E_2 \ldots\ldots E_m A F_1 \ldots F_t$

$$= \begin{pmatrix} I_r & O \\ O & O \end{pmatrix}, \quad r < n, \text{ by theorem 4.11.1.}$$

Taking determinants of both sides we get
$|E_1| \ldots\ldots |E_m||A||F_1 \ldots\ldots |F_t| = 0.$

Since all the $|E_i|$ and $|F_j|$ are non-zero we deduce that $|A| = 0$. This contradiction shows that rank $A \nless n$. Hence rank $A = n$.

Method for Finding A^{-1}. Let A be an $n \times n$ matrix over the field F. Let $|A| \neq 0$. Hence A^{-1} exists.
Reduce the $n \times 2n$ array $A \mid I_n$ to the array $I_n \mid B$, for some $n \times n$ matrix B. This can be done according to theorem 4.11.1, since rank $A = n$.

Moreover since rank $A = n$, the reader is left to check that only row operations are required.

Hence $E_1 \; E_2 \; E_3 \ldots\ldots E_m A = I_n$ for suitable elementary matrices $E_1, E_2, \ldots\ldots, E_m$.

Thus $A^{-1} = E_1 \, E_2 \, E_3 \ldots\ldots E_m = E_1 \, E_2 \, E_3 \ldots\ldots E_m I_n = B$.

This method is often called the **exchange method** or the method by row operations.

EXAMPLE 4.11.2 Find the inverse of the matrix

$$A = \begin{pmatrix} 1 & 5 & 2 \\ 1 & 1 & 7 \\ 0 & -3 & 4 \end{pmatrix}.$$

Solution. Let ⓘ denote row i. Then we have the following sequence:

$$\left.\begin{array}{rrr|rrr} 1 & 5 & 2 & 1 & 0 & 0 \\ 1 & 1 & 7 & 0 & 1 & 0 \\ 0 & -3 & 4 & 0 & 0 & 1 \end{array}\right. \xrightarrow{\text{② − ①}} \left.\begin{array}{rrr|rrr} 1 & 5 & 2 & 1 & 0 & 0 \\ 0 & -4 & 5 & -1 & 1 & 0 \\ 0 & -3 & 4 & 0 & 0 & 1 \end{array}\right.$$

$$\xrightarrow{\text{② − ③}} \left.\begin{array}{rrr|rrr} 1 & 5 & 2 & 1 & 0 & 0 \\ 0 & -1 & 1 & -1 & 1 & -1 \\ 0 & -3 & 4 & 0 & 0 & 1 \end{array}\right. \xrightarrow{\text{③ − 3 x ②}} \left.\begin{array}{rrr|rrr} 1 & 5 & 2 & 1 & 0 & 0 \\ 0 & -1 & 1 & -1 & 1 & -1 \\ 0 & 0 & 1 & 3 & -3 & 4 \end{array}\right.$$

$$\xrightarrow{\text{② − ③}} \left.\begin{array}{rrr|rrr} 1 & 5 & 2 & 1 & 0 & 0 \\ 0 & -1 & 0 & -4 & 4 & -5 \\ 0 & 0 & 1 & 3 & -3 & 4 \end{array}\right. \xrightarrow{\text{① − 2 x ③}} \left.\begin{array}{rrr|rrr} 1 & 5 & 0 & -5 & 6 & -8 \\ 0 & -1 & 0 & -4 & 4 & -5 \\ 0 & 0 & 1 & 3 & -3 & 4 \end{array}\right.$$

$$\xrightarrow{\text{① + 5 x ②}} \left.\begin{array}{rrr|rrr} 1 & 0 & 0 & -25 & 26 & -33 \\ 0 & -1 & 0 & -4 & 4 & -5 \\ 0 & 0 & 1 & 3 & -3 & 4 \end{array}\right. \xrightarrow{\text{−1 x ②}} \left.\begin{array}{rrr|rrr} 1 & 0 & 0 & -25 & 26 & -33 \\ 0 & 1 & 0 & 4 & -4 & 5 \\ 0 & 0 & 1 & 3 & -3 & 4 \end{array}\right.$$

$$\text{Thus} \quad A^{-1} = \begin{pmatrix} 1 & 5 & 2 \\ 1 & 1 & 7 \\ 0 & -3 & 4 \end{pmatrix}^{-1} = \begin{pmatrix} -25 & 26 & -33 \\ 4 & -4 & 5 \\ 3 & -3 & 4 \end{pmatrix}.$$

Direct calculation shows that this does satisfy

$$A A^{-1} = I_3 = A^{-1} A.$$

4.12 DETERMINANTAL RANK

Let A be an $m \times n$ matrix of rank r. Then A has r linearly independent rows. Thus we can obtain an $r \times n$ submatrix A_1 whose rows are linearly independent. Hence the rank of $A_1 = r$.

Thus A_1 has r linearly independent columns. Taking these columns we get an $r \times r$ submatrix (minor) A_2 of A_1 such that the rank of A_2 is r.

But then $|A_2| \neq 0$. Thus if rank $A = r$, then A has an $r \times r$ minor whose determinant is non-zero. (See section 4.5 for the definition of minors, or submatrices as they are also called).

Conversely suppose that A is an $m \times n$ matrix with a minor B of order p, that is $p \times p$, whose determinant is non-zero. Since $|B| \neq 0$, the rows of B are linearly independent, hence the same set of p rows of the whole matrix must be linearly independent. Any linear dependence relation involving these p rows of the whole matrix would give a linear dependence relation for the corresponding p rows of B. Thus the rank of $A \geqq p$.

Now suppose that A has an $r \times r$ minor whose determinant does not vanish while every minor of higher order has zero determinant. Then rank $A \geqq r$ by the above. On the other hand, by the first part above, we have: $r + 1 >$ rank A.

Altogether we have the rank of $A = r$. We have proved

THEOREM 4.12.1 Let A be an $m \times n$ matrix over the field F. Then the rank of A is the order of the minor of greatest order whose determinant does not vanish.

Often the order of the minor of greatest order whose determinant is non-zero is called the **determinantal rank** of A. The contents of theorems 4.10.1 and 4.12.1 may be expressed succinctly as:

column rank = row rank = determinantal rank.

EXAMPLE 4.12.1 Find the row, column, and determinantal ranks of

$$A = \begin{pmatrix} 2 & -3 & -1 & 1 \\ 3 & 4 & -4 & -3 \\ 0 & 17 & -5 & -9 \end{pmatrix} \quad \text{over the field } \mathbf{Q}.$$

Solution. We use elementary column operations, which do not affect the column rank, in order to reduce the columns to a form when the rank is seen by inspection.

Similar methods work for the row rank. We have the following sequences in each case.

$$\begin{pmatrix} 2 & -3 & -1 & 1 \\ 3 & 4 & -4 & -3 \\ 0 & 17 & -5 & -9 \end{pmatrix} \xrightarrow[\substack{\textcircled{1}-2\times\textcircled{4} \\ \textcircled{2}+3\times\textcircled{4} \\ \textcircled{3}+\textcircled{4}}]{} \begin{pmatrix} 0 & 0 & 0 & 1 \\ 9 & -5 & -7 & -3 \\ 18 & -10 & -14 & -9 \end{pmatrix}$$

$$\xrightarrow{3\times\textcircled{1}} \begin{pmatrix} 0 & 0 & 0 & 1 \\ 27 & -5 & -7 & -3 \\ 54 & -10 & -14 & -9 \end{pmatrix}$$

$$\xrightarrow{\textcircled{1}+4\times\textcircled{2}+\textcircled{3}} \begin{pmatrix} 0 & 0 & 0 & 1 \\ 0 & -5 & -7 & -3 \\ 0 & -10 & -14 & -9 \end{pmatrix} \xrightarrow{\textcircled{2}-\frac{5}{7}\times\textcircled{3}} \begin{pmatrix} 0 & 0 & 0 & 1 \\ 0 & 0 & -7 & -3 \\ 0 & 0 & -14 & -9 \end{pmatrix}.$$

Thus the column rank is 2.

$$\begin{pmatrix} 2 & -3 & -1 & 1 \\ 3 & 4 & -4 & -3 \\ 0 & 17 & -5 & -9 \end{pmatrix} \xrightarrow{\textcircled{2}-\textcircled{1}} \begin{pmatrix} 2 & -3 & -1 & 1 \\ 1 & 7 & -3 & -4 \\ 0 & 17 & -5 & -9 \end{pmatrix}$$

$$\xrightarrow{\textcircled{1}-2\times\textcircled{2}} \begin{pmatrix} 0 & -17 & 5 & 9 \\ 1 & 7 & -3 & -4 \\ 0 & 17 & -5 & -9 \end{pmatrix}$$

$$\xrightarrow[\textcircled{1}\text{ and }\textcircled{2}]{\text{Interchange}} \begin{pmatrix} 1 & 7 & -3 & -4 \\ 0 & -17 & 5 & 9 \\ 0 & 17 & -5 & -9 \end{pmatrix} \xrightarrow{\textcircled{3}-\textcircled{2}} \begin{pmatrix} 1 & 7 & -3 & -4 \\ 0 & -17 & 5 & 9 \\ 0 & 0 & 0 & 0 \end{pmatrix}$$

Thus the row rank is 2.

There are four minors of A of order 3 with determinants as follows.

$$\begin{vmatrix} 2 & -3 & -1 \\ 3 & 4 & -4 \\ 0 & 17 & -5 \end{vmatrix} = 0, \qquad \begin{vmatrix} 2 & -3 & 1 \\ 3 & 4 & -3 \\ 0 & 17 & -9 \end{vmatrix} = 0,$$

$$\begin{vmatrix} 2 & -1 & 1 \\ 3 & -4 & -3 \\ 0 & -5 & -9 \end{vmatrix} = 0 \qquad \begin{vmatrix} -3 & -1 & 1 \\ 4 & -4 & -3 \\ 17 & -5 & -9 \end{vmatrix} = 0$$

Thus the determinantal rank of A is < 3. On the other hand

$$\begin{vmatrix} 2 & -3 \\ 3 & 4 \end{vmatrix} = 17 \neq 0.$$ Thus the determinantal rank of $A \geqq 2$.

Together the two inequalities show that the determinantal rank of A is 2.

4.13 THE SOLUTION OF LINEAR EQUATIONS

We illustrate by examples a method of solving linear equations by reducing them to an echelon form by means of elementary row operations. This technique is sometimes called Gaussian reduction.

EXAMPLE 4.13.1 We want to solve the following system of simultaneous linear equations over the rational field **Q**.

$$\begin{aligned} x_1 + x_2 + 2x_3 + x_4 &= 5 \\ 3x_1 + 2x_2 - x_3 + 3x_4 &= 6 \\ 4x_1 + 3x_2 + x_3 + 4x_4 &= 11 \\ 2x_1 + x_2 - 3x_3 + 2x_4 &= 11 \end{aligned}$$

We ignore the variables x_1, x_2, x_3, x_4 and write the array of coefficients together with the right hand sides of the equations as follows.

$$\left.\begin{array}{cccc|c} 1 & 1 & 2 & 1 & 5 \\ 3 & 2 & -1 & 3 & 6 \\ 4 & 3 & 1 & 4 & 11 \\ 2 & 1 & -3 & 2 & 1 \end{array}\right. .$$

Now we reduce this array to echelon form by a sequence of elementary row operations as we did to find the row rank of a matrix in Example 4.12.1.

We have the following sequence:

$$\left(\begin{array}{cccc|c} 1 & 1 & 2 & 1 & 5 \\ 3 & 2 & -1 & 3 & 6 \\ 4 & 3 & 1 & 4 & 11 \\ 2 & 1 & -3 & 2 & 1 \end{array}\right) \xrightarrow[\text{④} - 2 \times \text{①}]{\text{②} - 3 \times \text{①},\ \text{③} - 4 \times \text{①}} \left(\begin{array}{cccc|c} 1 & 1 & 2 & 1 & 5 \\ 0 & -1 & -7 & 0 & -9 \\ 0 & -1 & -7 & 0 & -9 \\ 0 & -1 & -7 & 0 & -9 \end{array}\right)$$

$$\xrightarrow[\text{④} - \text{②}]{\text{③} - \text{②}} \left(\begin{array}{cccc|c} 1 & 1 & 2 & 1 & 5 \\ 0 & -1 & -7 & 0 & -9 \\ 0 & 0 & 0 & 0 & 0 \\ 0 & 0 & 0 & 0 & 0 \end{array}\right).$$

Replacing the variables we have:

$$\begin{aligned} x_1 + x_2 + 2x_3 + x_4 &= 5 \\ -x_2 - 7x_3 &= -9. \end{aligned}$$

Take $x_4 = a$, where a is an arbitrary rational.

Take $x_3 = b$, where b is an arbitrary rational.

Then $-x_2 = 7b - 9$, hence $x_2 = 9 - 7b$.

Also $x_1 = 5 - a - 2b - 9 + 7b = -a + 5b - 4$.

We can express this in the column vector form:

$$\begin{pmatrix} x_1 \\ x_2 \\ x_3 \\ x_4 \end{pmatrix} = a \begin{pmatrix} -1 \\ 0 \\ 0 \\ 1 \end{pmatrix} + b \begin{pmatrix} 5 \\ -7 \\ 1 \\ 0 \end{pmatrix} + \begin{pmatrix} -4 \\ 9 \\ 0 \\ 0 \end{pmatrix}.$$

We note that:

$$\left\langle \begin{pmatrix} -1 \\ 0 \\ 0 \\ 1 \end{pmatrix}, \begin{pmatrix} 5 \\ -7 \\ 1 \\ 0 \end{pmatrix} \right\rangle$$

is a vector space any of whose vectors satisfy the given equations when

$$\begin{pmatrix} 0 \\ 0 \\ 0 \\ 0 \end{pmatrix}$$

replaces the right hand side. The modified system can be written $\boldsymbol{Ax} = \boldsymbol{0}$, where A is the matrix of coefficients. This vector space

$$\left\langle \begin{pmatrix} -1 \\ 0 \\ 0 \\ 1 \end{pmatrix}, \begin{pmatrix} 5 \\ -7 \\ 1 \\ 0 \end{pmatrix} \right\rangle$$

is called the solution space of $\boldsymbol{Ax} = \boldsymbol{0}$.

$$\boldsymbol{x} = \begin{pmatrix} -4 \\ 9 \\ 0 \\ 0 \end{pmatrix}$$

gives a **particular** solution of the system

$$\boldsymbol{Ax} = \begin{pmatrix} 5 \\ 6 \\ 11 \\ 1 \end{pmatrix}.$$

The general solution of

$$\boldsymbol{Ax} = \begin{pmatrix} 5 \\ 6 \\ 11 \\ 1 \end{pmatrix}$$

is then the **sum** of

$$\left\langle \begin{pmatrix} -1 \\ 0 \\ 0 \\ 1 \end{pmatrix}, \begin{pmatrix} 5 \\ -7 \\ 1 \\ 0 \end{pmatrix} \right\rangle \text{ and } \begin{pmatrix} -4 \\ 9 \\ 0 \\ 0 \end{pmatrix},$$

namely the solution **set**:

$$\left\{ a\begin{pmatrix}-1\\0\\0\\1\end{pmatrix} + b\begin{pmatrix}5\\-7\\1\\0\end{pmatrix} + \begin{pmatrix}-4\\9\\0\\0\end{pmatrix} \quad : a, b \in \mathbf{Q} \right\}.$$

(Compare this result with that for second order linear differential equations with constant coefficients whose general solution is the sum of the complementary function and the particular integral).

Note that if we take our field to be **C** instead of **Q**, then the solution space is just the vector space generated by the same vectors

$$\begin{pmatrix}-1\\0\\0\\1\end{pmatrix}, \quad \begin{pmatrix}5\\-7\\1\\0\end{pmatrix}$$

but over **C** instead of **Q**.

EXAMPLE 4.13.2 We solve the system:

$$\begin{aligned} x_1 + x_2 - x_3 - x_4 &= -1 \\ 3x_1 + 4x_2 - x_3 - 2x_4 &= 3 \\ x_1 + 2x_2 + x_3 &= 5. \end{aligned}$$

In shorthand form this is:

$$Ax = \begin{pmatrix}-1\\3\\5\end{pmatrix} \quad , \text{where}$$

A is the 3 x 4 matrix $\begin{pmatrix}1 & 1 & -1 & 1\\3 & 4 & -1 & -2\\1 & 2 & 1 & 0\end{pmatrix}$.

We have the following sequence to echelon form.

$$\left.\begin{array}{cccc|c} 1 & 1 & -1 & 1 & -1\\ 3 & 4 & -1 & -2 & 3\\ 1 & 2 & 1 & 0 & 5 \end{array}\right. \quad \xrightarrow[\text{③} - \text{①}]{\text{②} - 3 \times \text{①}} \quad \begin{array}{cccc|c} 1 & 1 & -1 & -1 & -1\\ 0 & 1 & 2 & 1 & 6\\ 0 & 1 & 2 & 1 & 6 \end{array}$$

$$\xrightarrow{③ - ②} \left(\begin{array}{cccc|c} 1 & 1 & -1 & -1 & -1 \\ 0 & 1 & 2 & 1 & 6 \\ 0 & 0 & 0 & 0 & 0 \end{array}\right)$$

Replacing the variables we get:

$$\begin{array}{ccccccccc} x_1 & + & x_2 & - & x_3 & - & x_4 & = & -1 \\ & & x_2 & + & 2x_3 & + & x_4 & = & 6 \end{array}$$

Take $x_4 = a$, an arbitrary element in the field F over which the equations are taken. Thus F may be **Q**, **R** or **C**.

Take $x_3 = b$, an arbitrary element of F.

Then $x_2 = 6 - a - 2b$ and $x_1 = -1 + a + b - 6 + a + 2b$
$= -7 + 2a + 3b$.

Hence $\begin{pmatrix} x_1 \\ x_2 \\ x_3 \\ x_4 \end{pmatrix} = a\begin{pmatrix} 2 \\ -1 \\ 0 \\ 1 \end{pmatrix} + b\begin{pmatrix} 3 \\ -2 \\ 1 \\ 0 \end{pmatrix} + \begin{pmatrix} -7 \\ 6 \\ 0 \\ 0 \end{pmatrix}$.

The solution space of $Ax = 0$ is:

$$\left\langle \begin{pmatrix} 2 \\ -1 \\ 0 \\ 1 \end{pmatrix}, \begin{pmatrix} 3 \\ -2 \\ 1 \\ 0 \end{pmatrix} \right\rangle.$$

A particular solution of $Ax = \begin{pmatrix} -1 \\ 3 \\ 5 \end{pmatrix}$ is: $\begin{pmatrix} -7 \\ 6 \\ 0 \\ 0 \end{pmatrix}$.

The general solution of $Ax = \begin{pmatrix} -1 \\ 3 \\ 5 \end{pmatrix}$ is:

$$\left\langle \begin{pmatrix} 2 \\ -1 \\ 0 \\ 1 \end{pmatrix}, \begin{pmatrix} 3 \\ -2 \\ 1 \\ 0 \end{pmatrix} \right\rangle + \begin{pmatrix} -7 \\ 6 \\ 0 \\ 0 \end{pmatrix} = \left\{ a\begin{pmatrix} 2 \\ -1 \\ 0 \\ 1 \end{pmatrix} + b\begin{pmatrix} 3 \\ -2 \\ 1 \\ 0 \end{pmatrix} + \begin{pmatrix} -7 \\ 6 \\ 0 \\ 0 \end{pmatrix} : a, b \in F \right\}.$$

4.14 LINEAR TRANSFORMATIONS

We are now going to look more closely at the concept of the linear transformation which we defined in DEFINITION 4.9.3. Then we will use our results to give insight into the solution space and set of a system of simultaneous linear equations.

We remind the reader that a linear transformation f is a function from a vector space V over the field F into a vector space W over the same field F, such that $f(v_1 + v_2) = f(v_1) + f(v_2)$ for all $v_1, v_2 \in V$ and $f(av) = af(v)$ for all $a \in F$ and $v \in V$.

Associated with a given linear transformation f are two sets, one a subset of V and one a subset of W. These play a vital role in the theory.

DEFINITION 4.14.1 The **kernel of** f written Ker f is the subset of V defined by $\{v \in V : f(v) = 0\}$.

The **image of** f is the subset of W defined by $\text{Im}\, f = \{w \in W : w = f(v)$ for **some** $v \in V\}$.

The natural expectation is that Ker f is a subspace of V and Im f is a subspace of W. Before checking this we need a few elementary properties of linear transformations. We give these as

THEOREM 4.14.1 Let f be a linear transformation from the vector space V over the field F into the vector space W over F. Let 0 denote the zero of **both** vector spaces. Then

(1) $f(0) = 0$

(2) $f(-v) = -f(v)$ for all $v \in V$

Note that this is a special case of a similar result for groups. In fact the above holds because V and W are abelian groups.

PROOF

(1) $f(v + 0) = f(v) + f(0)$

Thus $f(v) = f(v) + f(0)$.

Adding the additive inverse of $f(v)$ to both sides of this equation cancels $f(v)$ and leaves us with $0 = f(0)$ as required.

(2) $f(v + (-v)) = f(v) + f(-v)$

Thus $f(0) = f(v) + f(-v)$

Hence, by (1) above, $0 = f(v) + f(-v)$.

Thus $-f(v) = -f(v) + (f(v) + f(-v))$

By associative rule we have:

$$-f(v) = (-f(v) + f(v)) + f(-v)$$

Finally $-f(v) = f(-v)$ as required.

THEOREM 4.14.2 Let f be a linear transformation from V into W. Then Ker f is a subspace of V and Im f is a subspace of W.
Moreover dim $V =$ dim Ker f + dim Im f.

PROOF

Let v_1 and v_2 be in Ker f.

Then $f(v_1) = f(v_2) = 0$.

Thus $f(v_1 + v_2) = f(v_1) + f(v_2) = 0 + 0 = 0$.

Hence $v_1 + v_2 \in \text{Ker } f$.

Also if $v \in$ Ker f and $a \in F$, the field over which the vector spaces are defined, then $f(av) = af(v) = a0 = 0$.

Thus $av \in \text{Ker } f$.

These two results show that Ker f is a subspace of V.

Now let $w_1, w_2 \in \text{Im } f$.

Then $w_1 = f(v_1)$ and $w_2 = f(v_2)$

for some $v_1, v_2 \in V$.

Thus $w_1 + w_2 = f(v_1) + f(v_2) = f(v_1 + v_2)$.

Hence $w_1 + w_2 \in \text{Im } f$.

Let $w \in$ Im f and $a \in F$.

Then $w = f(v)$ for some $v \in V$.

Thus $\quad aw = af(v) = f(av).$

Hence $\quad aw \in \operatorname{Im} f.$

Thus $\operatorname{Im} f$ is a subspace of W.

To prove that $\dim V = \dim \operatorname{Ker} f + \dim \operatorname{Im} f$, we take $\{v_1, v_2, \ldots\ldots, v_r\}$ as a basis of $\operatorname{Ker} f$ and extend it to a basis $\{v_1, v_2, \ldots\ldots, v_r, v_{r+1}, \ldots\ldots, v_n\}$ of V according to theorem 4.8.3.

Let w be **any** vector in $\operatorname{Im} f$. Then $w = f(v)$ for some $v \in V$.

Let $\quad v = a_1 v_1 + a_2 v_2 + \ldots\ldots + a_n v_n, a_i \in F.$

Then

$$w = f(v) = a_1 f(v_1) + \ldots\ldots + a_r f(v_r) + a_{r+1} f(v_{r+1}) + \ldots\ldots + a_n f(v_n)$$

$$= a_{r+1} f(v_{r+1}) + \ldots\ldots + a_n f(v_n),$$

because $f(v_1) = \ldots\ldots = f(v_r) = 0.$

Hence $\operatorname{Im} f$ is generated by $f(v_{r+1}) \ldots\ldots, f(v_n)$.

Now suppose that $\quad a_{r+1} f(v_{r+1}) + \ldots\ldots + a_n f(v_n) = 0$

for some $\quad a_j \in F.$

Then $\quad f(a_{r+1} v_{r+1} + \ldots\ldots + a_n v_n) = 0.$

Thus $\quad a_{r+1} v_{r+1} + \ldots\ldots + a_n v_n \in \operatorname{Ker} f.$

But $\quad \{v_1, v_2, \ldots\ldots, v_r\}$ is a basis for $\operatorname{Ker} f$.

Hence $\quad a_{r+1} v_{r+1} + \ldots\ldots + a_n v_n = b_1 v_1 + \ldots\ldots + b_r v_r$

for some $\quad b_i \in F.$

However this leads to a linear dependence relation for $\{v_1, v_2, \ldots\ldots, v_r, \ldots\ldots, v_n\}$ which contradicts the fact that $\{v_1, v_2, \ldots\ldots, v_n\}$ is a basis for V.

Hence $\{f(v_{r+1}), \ldots\ldots, f(v_n)\}$ is a linearly independent set. Altogether $\{f(v_{r+1}), \ldots\ldots, f(v_n)\}$ is a basis for $\operatorname{Im} f$.

Thus $\quad \dim V = n = r + (n - r) = \dim \operatorname{Ker} f + \dim \operatorname{Im} f.$

4.15 REPRESENTATION OF A LINEAR TRANSFORMATION BY A MATRIX

Let f be a linear transformation from the vector space V into the vector space W.

We suppose that both vector spaces are over the same field F. Let dim $V = n$ and dim $W = m$.

We choose a basis $\{v_1, v_2, \ldots\ldots, v_n\}$ for V and a basis $\{w_1, w_2, \ldots\ldots, w_m\}$ for W.

For each basis vector v_i, $f(v_i)$ is a vector of W and therefore can be expressed in the form $b_1w_1 + b_2w_2 + \ldots\ldots + b_mw_m$, where the b_j are in F.

Now the b_j will change as the v_i change. To reflect this we use a slightly different notation and write:

$$f(v_i) = \sum_{j=1}^{m} a_{ji} w_j, \text{ where } i = 1, 2, \ldots\ldots, n;$$

$$j = 1, 2, \ldots\ldots, m; \text{ and the } a_{ji} \text{ lie in } F.$$

This defines a matrix $A = (a_{ji})$. We say that the linear transformation f is **represented** by A.

It is easy for the reader to check that if f is represented by A and h is represented by B then $f + h$ is represented by $A + B$. Here $f + h$ is defined by $(f + h)(v) = f(v) + h(v)$, where $v \in V$. It is a linear transformation.

If $f : V \longrightarrow W$ and $g : W \longrightarrow U$ then the composition of f and g, written $g \circ f$, is a linear transformation from V into U. We show that $g \circ f$ is represented by BA when B represents g and A represents f.

Let V and W have bases as above. Let U have basis $\{u_1, u_2, \ldots\ldots, u_p\}$. We have

$$f(v_i) = \sum_{j=1}^{m} a_{ji} w_j \text{ and } g(w_j) = \sum_{s=1}^{p} b_{sj} u_s.$$

$$\text{Then } (g \circ f)(v_i) = g\left(\sum_{j=1}^{m} a_{ji} w_j\right) = \sum_{j=1}^{m} a_{ji}\, g(w_j) =$$

$$= \sum_{j=1}^{m} a_{ji} \sum_{s=1}^{p} b_{sj} u_s = \sum_{s=1}^{p} \left(\sum_{j=1}^{m} b_{sj} a_{ji}\right) u_s$$

$$= \sum_{s=1}^{p} c_{si} u_s, \text{ where } c_{si} = \sum_{j=1}^{m} b_{sj} a_{ji}.$$

$$\text{Thus} \quad C = (c_{si}) = BA.$$

Let f be a linear transformation from the vector space V into the vector space W. Let $a \in F$, the field over which both V and W are defined.

We define the multiplication of f by the scalar a to be the linear transformation af satisfying $(af)(v) = a(f(v))$ for all $v \in V$.

We leave the reader to check that this is indeed a linear transformation and that af is represented by the matrix aA, where A represents f.

The reader is also invited to check that the addition and multiplication by a scalar defined above turns the set of **all** linear transformations from V into W into a vector space over F. This vector space is denoted by $\text{Hom}_F(V, W)$. The Hom stands for homomorphisms which linear transformations are.

If we take $V = W$ then the composition $g \circ f$ defines a multiplication on $\text{Hom}_F(V, V)$ relative to which $\text{Hom}_F(V, V)$ is a **ring**. But $\text{Hom}_F(V, V)$ is not just an abelian group, it is a vector space. The scalar multiplication interacts with the ring multiplication according to the rule:

$$a(g \circ f) = (ag) \circ f = g \circ (af).$$

Any structure like this is called an **algebra over** F.

The representation of linear transformations by matrices also sets up an isomorphism between the vector space of all $m \times n$ matrices over F and $\text{Hom}_F(V, W)$, and between the algebras $M_n(F)$ and $\text{Hom}_F(V, V)$ over F. The latter is an isomorphism of both vector spaces and rings.

These statements are all easy, though perhaps a little tedious, to check. They all involve the checking of the various axioms, and defining conditions.

EXAMPLE 4.15.1

Let $\quad V = R^4,\ W = R^2.$

We define the function

$$f : R^4 \longrightarrow R^2 \text{ by } f((x_1, x_2, x_3, x_4)) = (x_1 + x_2, x_3 - x_4).$$

The reader is invited to carry out the routine check that this is a linear transformation.

We find the matrix representing f relative to the bases

$$\{(1, 0, 1, 0), (0, 1, 0, 1), (1, 0, 0, 0), (0, 0, 0, 1)\}$$

of R^4 and $\{(1, 1), (0, 1)\}$ of R^2.

We have:

$$f((1, 0, 1, 0)) = (1, 1) = 1.(1, 1) + 0.(0, 1)$$

$$f((0, 1, 0, 1)) = (1, -1) = 1.(1, 1) + (-2).(0, 1)$$

$$f((1, 0, 0, 0)) = (1, 0) = 1.(1, 1) + (-1).(0, 1)$$

$$f((0,0,0,1)) = (0,-1) = 0.(1,1) + (-1).(0,1)$$

Thus the matrix representing f relative to the given bases is

$$A = \begin{pmatrix} 1 & 1 & 1 & 0 \\ 0 & -2 & -1 & -1 \end{pmatrix}$$

4.16 REPRESENTATION OF ABSTRACT SYSTEM BY F^n, F^m AND A MATRIX A.

Suppose we have an abstract system consisting of vector spaces V and W over the field F and a linear transformation $f: V \longrightarrow W$. Let dim $V = n$ and dim $W = m$. By theorem 4.9.4 we can replace V by F^n and W by F^m. As we have just seen f can be represented by an $m \times n$ matrix A relative to bases $\{v_1, v_2, \ldots\ldots, v_n\}$ of V and $\{w_1, w_2, \ldots\ldots, w_m\}$ of W.

Define the function $f_A : F^n \longrightarrow F^m$ by

$$f_A((x_1, x_2, \ldots\ldots, x_n)) = (x_1, x_2, \ldots\ldots, x_n)A', \text{where the } x_i \in F.$$

From $f_A((x_1, x_2, x_3, \ldots, x_n) + (x_1', x_2', \ldots, x_n')) =$

$$= f_A((x_1 + x_1', \ldots\ldots, x_n + x_n')) =$$

$$= (x_1 + x_1', \ldots, x_n + x_n')A' = (x_1, \ldots, x_n)A' + (x_1', x_2', \ldots, x_n')A' =$$

$$= f_A((x_1, x_2, \ldots, x_n)) + f_A((x_1', x_2', \ldots, x_n')), \text{ and, for } a \in F,$$

$$f_A(a(x_1, x_2, \ldots, x_n)) = f_A((ax_1, ax_2, \ldots\ldots, ax_n)) =$$

$$= (ax_1, ax_2, \ldots, ax_n)A' = a(x_1, x_2, \ldots, x_n)A'$$

$$= a f_A((x_1, x_2, \ldots, x_n)),$$

we deduce that f_A is a linear transformation.

Let $f(v) = w$, where $v = \sum_{i=1}^{n} x_i v_i$ and $w = \sum_{j=1}^{m} y_j w_j$.

Then v in V corresponds to $(x_1, x_2, \ldots\ldots, x_n)$ in F^n and w in W corresponds to $(y_1, y_2, \ldots\ldots, y_m)$ in F^m.

Also $f(v) = \sum_{i=1}^{n} x_i f(v_i) = \sum_{i=1}^{n} x_i \sum_{j=1}^{m} a_{ji} w_j = \sum_{j=1}^{m} (\sum_{i=1}^{n} a_{ji} x_i) w_j =$

$$= w = \sum_{j=1}^{m} y_j w_j.$$

Now $\{w_1, w_2, \ldots\ldots, w_m\}$ is a basis.

Thus $\sum_{i=1}^{n} a_{ji} x_i = y_j$.

Hence $\begin{pmatrix} y_1 \\ \cdot \\ \cdot \\ y_m \end{pmatrix} = A \begin{pmatrix} x_1 \\ \cdot \\ \cdot \\ x_n \end{pmatrix} = A\mathbf{x}$, say.

Thus $(y_1, y_2, \ldots\ldots, y_m) = (A\mathbf{x})' = (x_1, x_2, \ldots\ldots, x_n) A'$.

Hence $(y_1, y_2, \ldots\ldots, y_m) = f_A((x_1, x_2, \ldots\ldots, x_n))$.

This means that the abstract system of V, W, and f can be replaced by the 'concrete' system F^n, F^m, and f_A, with the latter corresponding to f.

The diagram 4.16.1 may help the reader to visualise the situation.

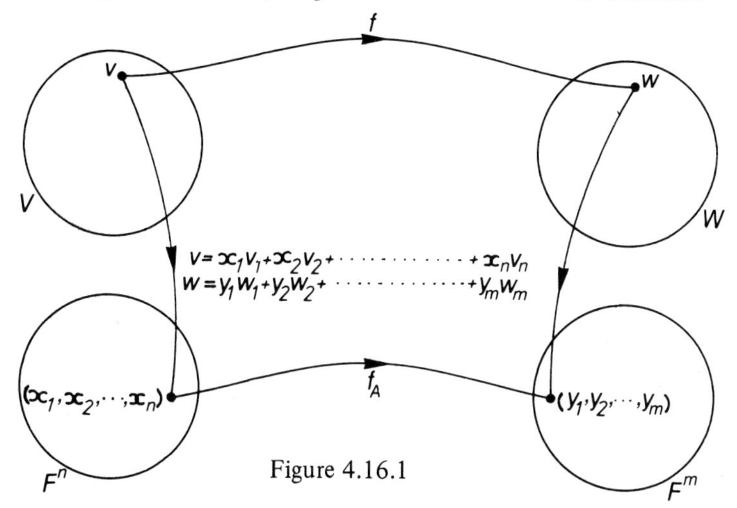

Figure 4.16.1

Alternatively we may take F_c^n and F_c^m to be the vector spaces of $n \times 1$ matrices, or column vectors.

Then we have:

$$f_A : F_c^n \longrightarrow F_c^m \text{ given by } f_A(\mathbf{x}) = A\mathbf{x}, \text{ where } \mathbf{x} = \begin{pmatrix} x_1 \\ x_2 \\ \cdot \\ \cdot \\ x_n \end{pmatrix}.$$

As above f_A is a linear transformation and f corresponds to f_A in the 'concrete' model.

Application to Rank of Matrix. We apply some of our ideas to matrices in order to prove some theorems about rank. These can be quite tedious to prove just by matrix manipulation.

Let $f_A : F_c^n \longrightarrow F_c^m$, where $f_A(x) = Ax$, $x = \begin{pmatrix} x_1 \\ \cdot \\ \cdot \\ \cdot \\ x_n \end{pmatrix}$, and

$$A = (a_{ij}), \text{ an } m \times n \text{ matrix.}$$

We denote the ith column of A by

$$a_i = \begin{pmatrix} a_{1i} \\ \cdot \\ \cdot \\ \cdot \\ a_{mi} \end{pmatrix}. \text{ Then } Ax = a_1 x_1 + \ldots + a_n x_n.$$

Thus the image of f_A is the vector space $\langle a_1, a_2, \ldots\ldots, a_n \rangle$.

But the column rank of A is the dimension of the column space of A, which is $\langle a_1, a_2, \ldots\ldots, a_n \rangle$. Thus the rank of $A = \dim \operatorname{Im} f_A$.

If V and W are abstract vector spaces with a linear transformation $f : V \longrightarrow W$ represented by the concrete system F_c^n, F_c^m, and f_A as above, then $\operatorname{Im} f$ must be isomorphic to $\operatorname{Im} f_A$. In particular $\dim \operatorname{Im} f_A = \dim \operatorname{Im} f$. The latter is often called the **rank of** f. The above considerations show that the rank of f is equal to the rank of A, the matrix representing f.

Note that f_A is an isomorphism if and only if $n = m$ and there exists a linear transformation g such that $g \circ f_A$ and $f_A \circ g$ are both identity transformations. Representing the linear transformation g by a matrix B, we see that f_A is an isomorphism if and only if $BA = I_n = AB$, that is if and only if A has an inverse.

We can now state and prove the following result on the ranks of products of matrices.

THEOREM 4.16.1 Let A be an $m \times n$ matrix over the field F. Let T be a **non-singular** $n \times n$ matrix over F, and let R be a **non-singular** $m \times m$ matrix over F. Then rank $(A\,T)$ = rank A = rank $(R\,A)$.

PROOF

f_T is a linear transformation from F_c^n into F_c^n and f_A is a linear transformation from F_c^n into F_c^m. From the above f_T is an isomorphism.

Also $\quad (f_A \circ f_T)\boldsymbol{x} = (A\,(T\,(\boldsymbol{x}))) = (A\,T)(\boldsymbol{x}) = f_{AT}\,(\boldsymbol{x}),$

for all $x \in F_c^n$.

Thus $f_A \circ f_T = f_{AT}$. These results enable us to say that : dim Im f_A = dim Im $(f_A \circ f_T)$ = dim Im f_{AT}.

From our earlier discussion in this section we deduce that : rank A = rank $(A\ T)$.

Figure 4.16.2 shows the above in a visual form.

In a similar way, using the linear transformations

$$f_A : F_c^n \longrightarrow F_c^m \text{ and } f_R : F_c^m \longrightarrow F_c^m,$$

as shown in Figure 4.16.3, we deduce that : rank A = rank $(R\,A)$.

This general technique can be applied in other situations of a similar kind.

Linear Equations. Finally we make use of the above ideas to gain insight into the structure of the solution space and set of a system of simultaneous linear equations.

Let A be an $m \times n$ matrix over the field F. Then a system of m linear equations in n variables may be written in the form:

$$A\boldsymbol{x} = \boldsymbol{k}, \text{ where } \boldsymbol{x} = \begin{pmatrix} x_1 \\ \cdot \\ \cdot \\ \cdot \\ x_n \end{pmatrix} \text{ and } \boldsymbol{k} = \begin{pmatrix} k_1 \\ \cdot \\ \cdot \\ \cdot \\ k_m \end{pmatrix}.$$

First we consider the case when $\boldsymbol{k} = \boldsymbol{0}$. Let $f_A : F_c^n \longrightarrow F_c^m$ be the linear transformation defined by $f_A\,(\boldsymbol{x}) = A\,\boldsymbol{x}$, as used before in this section.

Then the kernel of f_A is precisely the solution space of the equations $A\,\boldsymbol{x} = \boldsymbol{0}$.

By theorem 4.14.2, dim F_c^n = dim Ker f_A + dim Im f_A. As found earlier in this section, dim Im f_A = rank A, and, since Ker f_A = solution space of $A\,\boldsymbol{x} = \boldsymbol{0}$, dim Ker f_A = dim (solution space). Hence n = dim (solution space) + rank A. We have proved

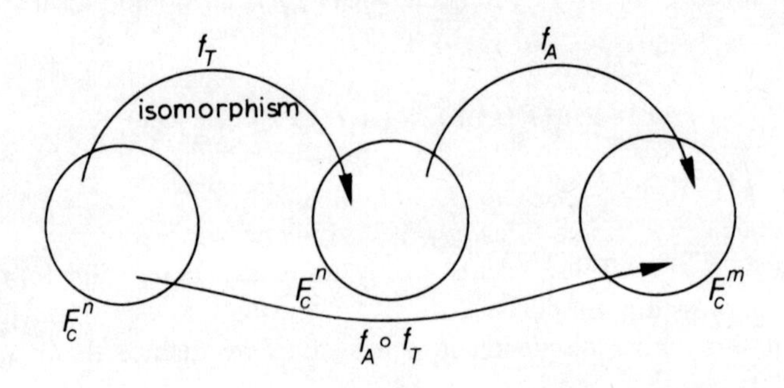

Figure 4.16.2

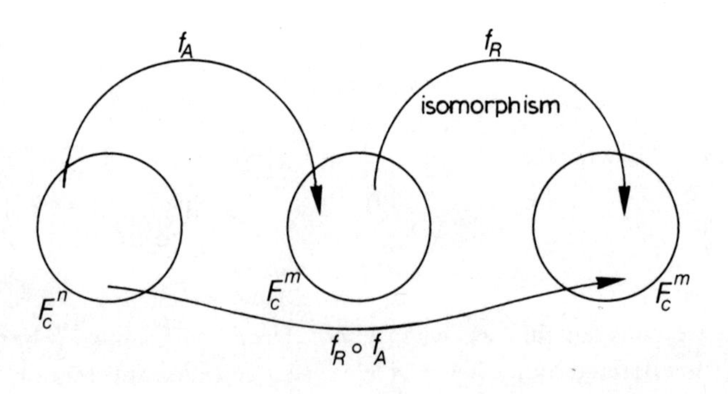

Figure 4.16.3

THEOREM 4.16.2 Let A be an $m \times n$ matrix over the field F. Let $A\ \boldsymbol{x} = \boldsymbol{0}$ be a system of simultaneous equations, where

$$\boldsymbol{x} = \begin{pmatrix} x_1 \\ \cdot \\ \cdot \\ \cdot \\ x_n \end{pmatrix}.$$

Then the dimension of the solution space of the system $A\ \boldsymbol{x} = \boldsymbol{0}$ is:

$$n - \operatorname{rank} A.$$

Now rank $A \leqq n$. Thus $n - \operatorname{rank} A \geqq 0$.
Hence there is a non-zero solution of $A\,\boldsymbol{x} = \boldsymbol{0}$ if and only if $n >$ rank A.

If $m < n$, then rank $A \leqq m < n$. Thus, if $m < n$, there is always a non-zero solution of $A\boldsymbol{x} = \boldsymbol{0}$.

If $m = n$, $A\ \boldsymbol{x} = \boldsymbol{0}$ has a non-zero solution if and only if $|A| = 0$. (Using determinantal rank).

This result is important enough to state again as

THEOREM 4.16.3 Let A be an $n \times n$ matrix over the field F. Then the system of n simultaneous equations in n unknowns $A\ \boldsymbol{x} = \boldsymbol{0}$ has a non-zero solution if and only if $|A| = 0$.

Now let us turn to the general case of m linear equations in n variables: $A\,\boldsymbol{x} = \boldsymbol{k}$, where A is an $m \times n$ matrix,

$$\boldsymbol{x} = \begin{pmatrix} x_1 \\ x_2 \\ \cdot \\ \cdot \\ \cdot \\ x_n \end{pmatrix}, \text{ and } \boldsymbol{k} = \begin{pmatrix} k_1 \\ k_2 \\ \cdot \\ \cdot \\ \cdot \\ k_m \end{pmatrix}.$$

The solution set of such a system is:

$$\{\boldsymbol{v}_0 + \boldsymbol{v} : \boldsymbol{v} \in \operatorname{Ker} f_A\} = \boldsymbol{v}_0 + \operatorname{Ker} f_A, \text{ where } A\ \boldsymbol{v}_0 = \boldsymbol{k}, \text{ and } f_A(\boldsymbol{x}) = A\ \boldsymbol{x}.$$

To prove this, let S be the solution set of $A\ \boldsymbol{x} = \boldsymbol{k}$.

Let $\quad \boldsymbol{v} \in S$.

Then $\quad f_A(\boldsymbol{v} - \boldsymbol{v}_0) = f_A(\boldsymbol{v}) - f_A(\boldsymbol{v}_0) = \boldsymbol{k} - \boldsymbol{k} = \boldsymbol{0}$.

Thus $\quad \boldsymbol{v} - \boldsymbol{v}_0 \in \operatorname{Ker} f_A$.

Hence $\quad v \in v_0 + \text{Ker}\, f_A.$

Thus $\quad S \subset v_0 + \text{Ker}\, f_A.$

Now let $v \in v_0 + \text{Ker}\, f_A.$

Then $\quad v = v_0 + v'$, where $v' \in \text{Ker}\, f_A.$

Thus $\quad f_A(v) = f_A(v_0) + f_A(v') = f_A(v_0) = k.$

Hence $\quad v \in S.$

Thus $\quad v_0 + \text{Ker}\, f_A \subset S.$

The two inclusions together show that $S = v_0 + \text{Ker}\, f_A$.
Let us recall that:

$$\text{Im}\, f_A = \{A\mathbf{x} : \mathbf{x} \in F_c^n\} = \{a_1 x_1 + a_2 x_2 + \ldots\ldots + a_n x_n : x_i \in F\}$$

$$= \langle a_1, a_2, \ldots\ldots, a_n \rangle.$$

Now the solution set is non-empty (i.e. the equations have a solution) if and only if $\mathbf{k} \in \text{Im}\, f_A$. This is so if and only if $\mathbf{k} \in \langle a_1, a_2, \ldots\ldots, a_n \rangle$, which is so if and only if $\langle a_1, a_2, \ldots\ldots, a_n \rangle = \langle a_1, \ldots\ldots, a_n, \mathbf{k} \rangle$.

Now this latter is true if and only if

dim $\quad \langle a_1, a_2, \ldots\ldots, a_n \rangle = \dim \langle a_1, a_2, \ldots\ldots, a_n, \mathbf{k} \rangle,$

since $\quad \langle a_1, \ldots\ldots, a_n \rangle \subset \langle a_1, a_2, \ldots\ldots, a_n, \mathbf{k} \rangle.$

But $\dim \langle a_1, a_2, \ldots\ldots, a_n \rangle = \text{rank}\, A$ and

$\dim \langle a_1, a_2, \ldots\ldots, a_n, \mathbf{k} \rangle = \text{rank}\, (A, \mathbf{k})$, where

$(A, \mathbf{k})$ denotes the so-called **augmented matrix**

$$\begin{pmatrix} a_{11} & a_{12} \cdots\cdots & a_{1n} & k_1 \\ a_{21} & a_{22} \cdots\cdots & a_{2n} & k_2 \\ \cdot & \cdot & \cdot & \cdot \\ \cdot & \cdot & \cdot & \cdot \\ \cdot & \cdot & \cdot & \cdot \\ \cdot & \cdot & \cdot & \cdot \\ a_{m1} & a_{m2} \cdots\cdots & a_{mn} & k_m \end{pmatrix}$$

Thus there is at least one solution of $A\,\boldsymbol{x} = \boldsymbol{k}$ if and only if rank A = rank $(A, \boldsymbol{k})$.

Altogether we have proved

THEOREM 4.16.4 Let $A\,\boldsymbol{x} = \boldsymbol{k}$ be a system of m linear equations in n variables, where A is an $m \times n$ matrix over the field F,

$$\boldsymbol{x} = \begin{pmatrix} x_1 \\ \cdot \\ \cdot \\ \cdot \\ x_n \end{pmatrix}, \text{ and } \boldsymbol{k} = \begin{pmatrix} k_1 \\ \cdot \\ \cdot \\ \cdot \\ k_m \end{pmatrix}, \text{ with } k_i \in F.$$

Then the solution set of $A\,\boldsymbol{x} = \boldsymbol{k}$ is:

$$\boldsymbol{v}_0 + \operatorname{Ker} f_A, \text{ where } f_A(\boldsymbol{x}) = A\,\boldsymbol{x}, \text{ and } A\,\boldsymbol{v}_0 = \boldsymbol{k}.$$

This set is non-empty if and only if:

$$\operatorname{rank} A = \operatorname{rank}(A, \boldsymbol{k}).$$

4.17 CHANGE OF BASES

We have mentioned already that the matrix representing a linear transformation f from the vector space V into the vector space W depends on the bases chosen for V and W. In this section we investigate this dependence in some detail.

Let V be a vector space over the field F and let W be a vector space over the same field. Let f be a linear transformation from V into W.

Let $\{v_1, v_2, \ldots\ldots, v_n\}$ and $\{w_1, w_2, \ldots\ldots, w_m\}$ be the first or original bases for V and W respectively.

Let $\{v_1', v_2', \ldots\ldots, v_n'\}$ and $\{w_1', w_2', w_3', \ldots\ldots, w_m'\}$ be the two second or new bases.

We have $f(v_i) = \sum_{j=1}^{m} a_{ji}\, w_j$ defining the $m \times n$ matrix $A = (a_{ji})$

representing f relative to the first bases.

Now $f(v_i') = \sum_{j=1}^{m} b_{ji}\, w_j'$, defines the matrix $B = (b_{ji})$

representing f relative to the second bases.

Let $v_i' = \sum_{r=1}^{n} p_{ri}\, v_r$ and $w_j' = \sum_{s=1}^{m} a_{sj}\, w_s$

define the change from the first to the second bases.

Then $f(v_i') = \sum_{r=1}^{n} p_{ri} f(v_r) = \sum_{r=1}^{n} p_{ri} \sum_{j=1}^{m} a_{jr} w_j =$

$$= \sum_{j=1}^{m} \sum_{r=1}^{n} (a_{jr} p_{ri}) w_j.$$

Also we have $f(v_i') = \sum_{s=1}^{m} b_{si} w_s' = \sum_{s=1}^{m} b_{si} \sum_{j=1}^{m} q_{js} w_j =$

$$= \sum_{j=1}^{m} (\sum_{s=1}^{m} q_{js} b_{si}) w_j.$$

Equating the two expressions for $f(v_i')$ and remembering that $\{w_1, w_2, \ldots\ldots, w_m\}$ is a basis of W, we have:

(1) $\sum_{r=1}^{n} a_{jr} p_{ri} = \sum_{s=1}^{m} q_{js} b_{si}$, where $j = 1, 2, 3, \ldots\ldots, m$ and $i = 1, 2, 3, 4, \ldots\ldots, \ldots\ldots, n$.

Let $P = (p_{ri})$ and $Q = (q_{sj})$. Then P is the matrix associated with the change of basis of V and Q with the change of basis of W.

Thus equation (1) asserts that the element in the jth row and ith column of AP is equal to the element in the same position of the matrix QB. This is true for **all** j and i.

Hence $AP = QB$.

Now the change of basis from $\{v_1, v_2, \ldots\ldots, v_n\}$ to $\{v_1', v_2', \ldots\ldots, v_n'\}$ defines a linear transformation f_P from V into V according to:

$$f_P(v_i) = v_i', \text{ and if } v = x_1 v_1 + x_2 v_2 + \ldots\ldots + x_n v_n$$

is **any** vector in V then $f_P(v) = x_1 f_P(v_1) \ldots. + x_n f_P(v_n)$.

Since both $\{v_1, v_2, \ldots\ldots, v_n\}$ and $\{v_1', v_2', \ldots\ldots, v_n'\}$ are bases, f_P is an **isomorphism** of V with itself, sometimes called an **automorphism** of V.

Now it is immediate on inspection that the matrix representing f_P relative to the bases $\{v_1, v_2, \ldots\ldots v_n\}$ and $\{v_1, v_2, \ldots\ldots, v_n\}$ is none other than the $n \times n$ matrix P. Since f_P is an isomorphism, as argued in a similar situation just prior to theorem 4.16.1, P must have an inverse. Similarly, since Q represents the linear transformation f_q from W into W defined by the change of basis from $\{w_1, w_2, w_3, \ldots\ldots, w_m\}$ to $\{w_1', w_2', \ldots\ldots, w_m'\}$, Q has an inverse.

This means that from $AP = QB$ we can deduce that $B = Q^{-1}AP$, which gives us the required relation connecting the matrices A and B that both represent the linear transformation $f : V \longrightarrow W$ but relative to different pairs of bases.

In the particular case when $V=W, v_i=w_i$, and $v_i'=w_i'$, we have $P=Q$ and

$$B = P^{-1} A P.$$

By exploiting this change of basis, it is possible under certain circumstances to find a matrix P such that a given matrix A transforms to $B=P^{-1} A P$ with B having a particularly simple form, such as

$$\begin{pmatrix} b_{11} & 0 & \cdots & 0 \\ 0 & b_{22} & \cdots & \cdot \\ \cdot & \cdot & \ddots & \cdot \\ \cdot & \cdot & & \cdot \\ \cdot & \cdot & & \cdot \\ 0 & \cdot & \cdots & b_{nn} \end{pmatrix},$$

while retaining many of the properties of A. This kind of matrix with non-zero values only down the leading diagonal is called a **diagonal matrix**. In fact this theory occupies a considerable part of most 'second' courses in linear algebra. We shall return to this topic later, but in this introductory book we will avoid making full use of the above technique.

EXAMPLE 4.17.1

Let $\quad f : (x_1, x_2, x_3) \longrightarrow (x_1 - x_2, x_2 - x_3)$

be a function from $\mathbf{R}^3$ into $\mathbf{R}^2$.

(1) Show that f is a linear transformation.

(2) Find Ker f and dim Ker f.

(3) Find Im f and dim Im f.

(4) Find the matrix A representing f relative to the bases $\{(1,0,0), (0,1,1), (0,-1,0)\}$ of $\mathbf{R}^3$ and $\{(1,0), (0,1)\}$ of $\mathbf{R}^2$.

(5) Find the matrix B representing f relative to bases $\{(1,0,0), (1,0,1), (0,1,-1)\}$ of $\mathbf{R}^3$ and $\{(1,1), (1,-1)\}$ of $\mathbf{R}^2$.

(6) Find the ranks of A and B by at least two methods.

(7) Find matrices P and Q such that $B = Q^{-1} A P$.

Solution.

(1) $f((x_1, x_2, x_3) + (x_1', x_2', x_3')) = f((x_1 + x_1', x_2 + x_2', x_3 + x_3')) =$

$= (x_1 + x_1' - x_2 - x_2', x_2 + x_2' - x_3 - x_3')$

$= f((x_1, x_2, x_3)) + f((x_1', x_2', x_3'))$.

Also $f(a(x_1, x_2, x_3)) = af((x_1, x_2, x_3))$; $a \in \mathbf{R}$.

(2) For $(x_1, x_2, x_3) \in \operatorname{Ker} f$ we have $x_1 - x_2 = 0, x_2 - x_3 = 0$.

Thus $(x_1, x_2, x_3) = a(1,1,1)$.

Hence $\operatorname{Ker} f = \langle\langle(1,1,1)\rangle\rangle$.

Thus $\dim \operatorname{Ker} f = 1$.

(3) $\operatorname{Im} f = \langle f((1,0,0)), f((0,1,0)), f((0,0,1)) \rangle =$

$= \langle (1,0), (-1,1), (0,-1) \rangle = \langle (1,0), (0,-1) \rangle$

$= \mathbf{R}^2$.

Thus $\dim \operatorname{Im} f = 2$.

Alternatively, in this case, we could argue that

$\dim \mathbf{R}^3 = \dim \operatorname{Ker} f + \dim \operatorname{Im} f$.

Thus $3 = 1 + \dim \operatorname{Im} f$.

Hence $\dim \operatorname{Im} f = 2$.

Now $\operatorname{Im} f \subset \mathbf{R}^2$ of dimension 2.

Thus $\operatorname{Im} f$ must be the whole of $\mathbf{R}^2$.

(4) $f((1,0,0)) = (1,0) = 1 \cdot (1,0) + 0\,(0,1)$

$f((0,1,1)) = (-1,0) = -1 \cdot (1,0) + 0\,(0,1)$

$f((0,-1,0)) = (1,-1) = 1 \cdot (1,0) + (-1)(0,1)$

Thus f is represented by

$$A = \begin{pmatrix} 1 & -1 & 1 \\ 0 & 0 & -1 \end{pmatrix}.$$

(5) $f((1,0,0)) = (1,0) = \tfrac{1}{2}\cdot(1,1) + \tfrac{1}{2}\cdot(1,-1)$

$f((1,0,1)) = (1,-1) = 0\cdot(1,1) + 1\cdot(1,-1)$

$f((0,1,-1)) = (-1,2) = \tfrac{1}{2}\cdot(1,1) + (-\tfrac{3}{2})\cdot(1,-1)$

Thus $B = \begin{pmatrix} \tfrac{1}{2} & 0 & \tfrac{1}{2} \\ \tfrac{1}{2} & 1 & -\tfrac{3}{2} \end{pmatrix}$.

(6) Rank A = rank f = rank B. Now rank f = dim Im f = 2.

Also $A = \begin{pmatrix} 1 & -1 & 1 \\ 0 & 0 & -1 \end{pmatrix}$,

which is already echelon form. The rows are linearly independent by inspection. Thus rank A = 2 as a row rank.

$$B = \begin{pmatrix} \tfrac{1}{2} & 0 & \tfrac{1}{2} \\ \tfrac{1}{2} & 1 & -\tfrac{3}{2} \end{pmatrix}.$$

Rank $B \leqq 2$, because there are only 2 rows in B.

Now $\begin{vmatrix} \tfrac{1}{2} & 0 \\ \tfrac{1}{2} & 1 \end{vmatrix} \neq 0,$

hence determinantal rank of B is 2.

(7) $(1,0,0) = 1\cdot(1,0,0) + 0\,(0,1,1) + 0\,(0,-1,0)$

$(1,0,1) = 1\cdot(1,0,0) + 1\,(0,1,1) + 1\,(0,-1,0)$

$(0,1,-1) = 0\cdot(1,0,0) + (-1)(0,1,1) + (-2)(0,-1,0)$

gives the B basis in terms of the A basis, for $\mathbf{R}^3$.

Thus $P = \begin{pmatrix} 1 & 1 & 0 \\ 0 & 1 & -1 \\ 0 & 1 & -2 \end{pmatrix}$

$(1,1) = 1\cdot(1,0) + 1\cdot(0,1)$

$(1,-1) = 1\cdot(1,0) + (-1)(0,1)$, gives B basis in terms of A basis, for $\mathbf{R}^2$.

Thus $Q = \begin{pmatrix} 1 & 1 \\ 1 & -1 \end{pmatrix}$.

We have $Q^{-1} = \frac{1}{2}\begin{pmatrix} 1 & 1 \\ 1 & -1 \end{pmatrix}$.

$$\text{Hence}\quad Q^{-1}AP = \tfrac{1}{2}\begin{pmatrix} 1 & 1 \\ 1 & -1 \end{pmatrix}\begin{pmatrix} 1 & -1 & 1 \\ 0 & 0 & -1 \end{pmatrix}\begin{pmatrix} 1 & 1 & 0 \\ 0 & 1 & -1 \\ 0 & 1 & -2 \end{pmatrix}$$

$$= \tfrac{1}{2}\begin{pmatrix} 1 & 1 \\ 1 & -1 \end{pmatrix}\begin{pmatrix} 1 & 1 & -1 \\ 0 & -1 & 2 \end{pmatrix}$$

$$= \tfrac{1}{2}\begin{pmatrix} 1 & 0 & 1 \\ 1 & 2 & -3 \end{pmatrix}$$

$$= \begin{pmatrix} \frac{1}{2} & 0 & \frac{1}{2} \\ \frac{1}{2} & 1 & -\frac{3}{2} \end{pmatrix} = B, \text{ as required.}$$

EXERCISES

1. Let $A = \begin{pmatrix} 1 & 2 & -1 & 0 \\ 4 & 0 & 2 & 1 \\ 2 & -5 & 1 & 2 \end{pmatrix}$, $B = \begin{pmatrix} 2 & 1 & -1 \\ 0 & 1 & 0 \\ -1 & 2 & 1 \end{pmatrix}$,

$C = \begin{pmatrix} 0 & 1 & 1 \\ -1 & 2 & 0 \end{pmatrix}$, $D = \begin{pmatrix} -1 & 2 \\ 0 & 1 \\ 5 & 0 \end{pmatrix}$,

$E = \begin{pmatrix} 1 & 0 & 0 \\ 0 & -1 & 2 \\ -1 & 0 & 1 \end{pmatrix}$, $F = \begin{pmatrix} 2 & -2 \\ 0 & 1 \end{pmatrix}$.

Calculate the following **when possible**:

(a) $A + B$ (b) $B + E$ (c) $C + F$ (d) $E - B$ (e) $C + D$

(f) AB (g) BA (h) CA (i) AC (j) BC (k) CB

(l) BE (m) EB (n) AE (o) EA (p) CF

(q) FC (r) $F^2 = FF$ (s) F^3 (t) $C(E+B)$

(u) $CE + CB$ (v) $(DF)C$ (w) $D(FC)$ (x) $(DC)E$

(y) $D(CE)$ (z) $(CD)F$ (z′) $C(DF)$.

Check that $C(E+B) = CE + CB$, $(DF)C = D(FC)$, $(DC)E = D(CE)$ and $(CD)F = C(DF)$.

2. Explain why, in general, $(A+B)^2 \neq A^2 + 2AB + B^2$, $(A-B)^2 \neq A^2 - 2AB + B^2$ and $A^2 - B^2 \neq (A-B)(A+B)$.

3. A matrix M is said to be **idempotent** if $M^2 = M$.

(a) Show that $\begin{pmatrix} 2 & -2 & -4 \\ -1 & 3 & 4 \\ 1 & -2 & -3 \end{pmatrix}$ is indempotent.

(b) Let $AB = A$ and $BA = B$. Show that A and B are indempotent.

4. A matrix M is said to be **nilpotent** of **index** n if

$$M^n = \bigcirc \quad \text{and} \quad M^r \neq \bigcirc \quad \text{for } r < n, \quad \text{where} \quad \bigcirc \text{ is the null}$$

or zero matrix.

(a) Show that $\begin{pmatrix} 1 & 1 & 3 \\ 5 & 2 & 6 \\ -2 & -1 & -3 \end{pmatrix}$ is nilpotent of index 3.

(b) If M is nilpotent of index 2, show that

$$M(I \pm M)^n = M$$

for any positive integer n, where I is the appropriate identity matrix.

5. Evaluate:

(a) $\begin{vmatrix} 1 & 0 & 2 \\ 3 & 4 & 5 \\ 5 & 6 & 7 \end{vmatrix}$ (b) $\begin{vmatrix} 1 & 0 & 6 \\ 3 & 4 & 15 \\ 5 & 6 & 21 \end{vmatrix}$ (c) $\begin{vmatrix} 1 & 0 & 2 & 1 \\ 0 & 1 & -1 & 2 \\ 3 & 0 & 1 & -1 \\ 2 & 1 & 4 & 5 \end{vmatrix}$.

6. Show that $\begin{vmatrix} 1 & a & b+c \\ 1 & b & c+a \\ 1 & c & a+b \end{vmatrix} = 0.$

7. Express $\begin{vmatrix} 1 & a & a^2 \\ 1 & b & b^2 \\ 1 & c & c^2 \end{vmatrix}$ as a product of linear factors.

8. Show that $\begin{vmatrix} bc & a^2 & a^2 \\ b^2 & ca & b^2 \\ c^2 & c^2 & ab \end{vmatrix} = \begin{vmatrix} bc & ab & ca \\ ab & ca & bc \\ ca & bc & ab \end{vmatrix}$

without evaluating either determinant.

9. Find the inverse of the matrix $\begin{pmatrix} 2 & 4 & 1 \\ 1 & 1 & 1 \\ 2 & 3 & 1 \end{pmatrix}$,

hence solve the system of linear equations:

$$2x + 4y + z = 5$$

$$x + y + z = 6$$

$$2x + 3y + z = 6.$$

10. Find the inverses of the following matrices:

(a) $\begin{pmatrix} 5 & 15 & -10 \\ -2 & -2 & -4 \\ 3 & 4 & 1 \end{pmatrix}$ (b) $\begin{pmatrix} 3 & 2 & -1 \\ 1 & 7 & 5 \\ -1 & 0 & 1 \end{pmatrix}$ (c) $\begin{pmatrix} 1 & 5 & 2 \\ 1 & 1 & 7 \\ 0 & -3 & 4 \end{pmatrix}$.

11. Find whether or not the following subsets are **subspaces** of $\mathbf{R}^4$.

(a) The set of all vectors (x_1, x_2, x_3, x_4) for which $x_1 = x_4$.

(b) The set of all vectors for which $x_1 + x_2 = 0$.

(c) The set of all vectors for which $x_1 x_2 = 0$.

12. Let V be the set of all functions f from the interval $S = \{x \in \mathbf{R} : 0 \leqq x \leqq 1\}$ into $\mathbf{R}$. Define the addition of two functions f_1 and f_2 by $(f_1 + f_2)(x) = f_1(x) + f_2(x)$, $x \in S$ and scalar multiplication by $(af)(x) = a(f(x))$, where $x \in S$ and $a \in \mathbf{R}$. Show that V is a vector space over $\mathbf{R}$. Determine which of the following subsets of V are **subspaces** of V.

(a) The set of all functions f for which $f(1) = 0$.

(b) The set of all functions f for which $f(0) = f(1)$.

(c) The set of all functions f for which $f(0) + f(1) = 0$.

(d) The set of all functions f for which $f(x) \geqq 0, x \in S$.

(e) The set of all functions f for which $f(1/2) = (f(0) + f(1))/2$.

13. Find a basis for the subspace of $\mathbf{R}^3$ generated by the vectors (1,2,0), (–1,1,0), (1,1,0).

14. Find a basis of the space $\langle$ (1,–1,2,3), (1,0,1,0), (3,–2,5,7) $\rangle$ which includes the vector (1,1,0,–1).

15. Extend the set {(1,1,–1,1), (1,0,1,1), (1,2,1,1)} to a basis of $\mathbf{R}^4$.

16. Find the rank of the following matrices:

(a) $\begin{pmatrix} 2 & 0 & -1 & 4 \\ 1 & -2 & 0 & 3 \\ 1 & 6 & -2 & -1 \end{pmatrix}$ (b) $\begin{pmatrix} 3 & 1 & -1 & 2 \\ 2 & 0 & 1 & -1 \\ -1 & 1 & -3 & 4 \\ 4 & 2 & -3 & 5 \end{pmatrix}$

17. By using the 'exchange' method (method by row operations) find the inverse of the following matrices:

(a) $\begin{pmatrix} 1 & 2 & 3 \\ 2 & 4 & 5 \\ 3 & 5 & 6 \end{pmatrix}$ (b) $\begin{pmatrix} 1 & 3 & 3 \\ 1 & 4 & 3 \\ 1 & 3 & 4 \end{pmatrix}$ (c) $\begin{pmatrix} 1 & 0 & 0 \\ 0 & -1 & 2 \\ -1 & 0 & 1 \end{pmatrix}$

18. The following arrays represent systems of linear equations according to the notation:

$$\begin{aligned} a_{11}x_1 + a_{12}x_2 + a_{13}x_3 &= k_1 \\ a_{21}x_1 + a_{22}x_2 + a_{23}x_3 &= k_2 \\ a_{31}x_1 + a_{32}x_2 + a_{33}x_3 &= k_3 \end{aligned} \quad \text{is represented by} \quad \begin{array}{ccc|c} a_{11} & a_{12} & a_{13} & k_1 \\ a_{21} & a_{22} & a_{23} & k_2 \\ a_{31} & a_{32} & a_{33} & k_3 \end{array}$$

Find the complete solution in the cases where there is one.

(a) $\left[\begin{array}{ccc|c} 2 & 4 & 1 & 1 \\ 3 & 5 & 0 & 1 \\ 5 & 13 & 7 & 4 \end{array}\right]$ (b) $\left[\begin{array}{ccc|c} 9 & -6 & 12 & 0 \\ -12 & 8 & -16 & 0 \\ -7 & 10 & -13 & 0 \end{array}\right]$

(c) $\left[\begin{array}{ccc|c} 2 & 3 & 4 & 1 \\ 5 & 6 & 7 & 2 \\ 8 & 9 & 10 & 4 \end{array}\right]$

(d) $\left[\begin{array}{cccc|c} 2 & 1 & 3 & 5 & 6 \\ 3 & 2 & 4 & 6 & 8 \\ -1 & 3 & 2 & 7 & -3 \end{array}\right]$ (e) $\left[\begin{array}{cccc|c} 3 & 1 & -2 & -1 & 2 \\ 15 & 5 & 4 & 3 & 4 \\ 6 & 2 & 3 & 1 & 0 \end{array}\right]$

19. A function f from $\mathbf{R}^3$ into $\mathbf{R}^2$ is defined by

$$f((x_1, x_2, x_3)) = (x_1 + x_2 + x_3,\ x_1 - x_2 - x_3).$$

(a) Show that f is a linear transformation.

(b) Find the kernel of f (Ker f) and its dimension.

(c) Find the dimension of Im f and describe Im f.

(d) Find a matrix A representing f relative to the basis $\{(1,1,0), (1,-1,0), (0,1,2)\}$ of $\mathbf{R}^3$ and the basis $\{(2,-1), (1,2)\}$ of $\mathbf{R}^2$.

20. A function f from $\mathbf{R}^3$ into $\mathbf{R}^3$ is defined by

$$f((x, y, z)) = (x + y,\ y - z,\ x + z).$$

(a) Show that f is a linear transformation.

(b) Find Ker f and its dimension.

(c) Find Im f and verify that dim $\mathbf{R}^3$ = dim Ker f + dim Im f.

(d) Find a matrix A representing f relative to the basis $\{(1,0,0), (0,-1,2), (-1,0,1,)\}$ of $\mathbf{R}^3$.

(e) Find a matrix B representing f relative to the basis $\{ (1,1,0),(1,-1,0),(0,1,2) \}$ of $\mathbf{R}^3$.

(f) Find a matrix P such that $A = P^{-1} B P$.

(g) Find the rank of A, the rank of B, and the rank of P.

21. Find a linear transformation $f : \mathbf{R}^3 \to \mathbf{R}^4$ whose image is generated by the vectors (1,2,0,–4) and (2,0,–1,–3).

22. Let V be the vector space of all 2 x 2 matrices over $\mathbf{R}$ under matrix addition and the usual multiplication of a matrix by a scalar. Let $f : V \to V$ be the function defined by $f(A) = A\,M - M\,A$, where $A \in V$ and M is the fixed matrix

$$\begin{pmatrix} 1 & 2 \\ 0 & 3 \end{pmatrix} .$$

(a) Show that f is a linear transformation.

(b) Find a basis for Ker f.

(c) Find the dimension of Im f.

23. Let $f : \mathbf{R}^3 \to \mathbf{R}^2$ be defined by $f((x,y,z)) = (2x, y+z)$. Let $g : \mathbf{R}^3 \to \mathbf{R}^2$ be defined by $g((x,y,z)) = (x-z, y)$.

Find similar expressions to define the functions $f+g$, $3f$, $2f-5g$.

24. Let $\mathrm{Hom}_{\mathbf{R}}(\mathbf{R}^3, \mathbf{R}^2)$ denote the set of all linear transformations from $\mathbf{R}^3$ into $\mathbf{R}^2$. With addition and scalar multiplication as defined in section 4.15 $\mathrm{Hom}_{\mathbf{R}}(\mathbf{R}^3, \mathbf{R}^2)$ is a vector space over the field $\mathbf{R}$.

Let $f, g, h \in \mathrm{Hom}_{\mathbf{R}}(\mathbf{R}^3, \mathbf{R}^2)$ be defined by

$$f((x,y,z)) = (x+y+z,\ x+y),$$

$$g((x,y,z)) = (2x+z,\ x+y),$$

$$h((x,y,z)) = (2y, x).$$

Show that f, g, h are linearly independent as vectors in the vector space $\mathrm{Hom}_{\mathbf{R}}(\mathbf{R}^3, \mathbf{R}^2)$.

CHAPTER 5

Lines, Planes, Spheres, and Coordinate Transformations

5.1 INTRODUCTION

We assume that the reader has some background knowledge of vectors in three dimensional geometry such as is contained in *Analytic Geometry and Vectors* by J. Hunter (Blackie and Chambers 1972). The reader without this background is referred to this book.

We adopt as our frame of reference the usual Cartesian system of mutually perpendicular straight line axes with origin 0. We label the axes x, y, and z, or x_1, x_2, x_3, as happens to be convenient, and take unit vectors $\boldsymbol{i}, \boldsymbol{j}, \boldsymbol{k}$ from the origin directed along these axes in the usual order. Unless otherwise stated the set of axes will be right-handed, i.e. $\boldsymbol{k} = \boldsymbol{i} \wedge \boldsymbol{j}$, the vector product of $\boldsymbol{i}$ and $\boldsymbol{j}$. Figure 5.1.1 illustrates what we have in mind.

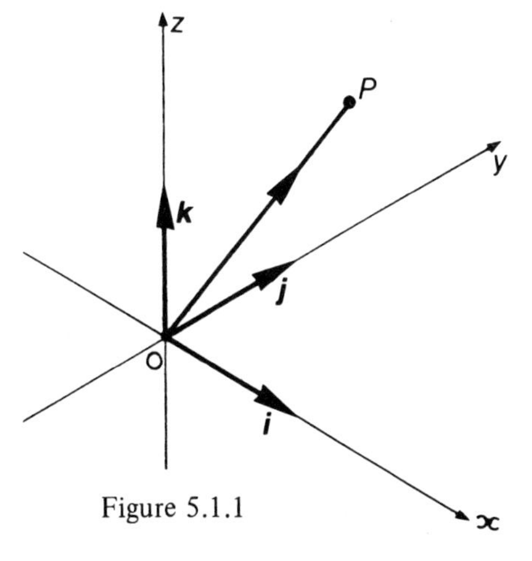

Figure 5.1.1

The point P with coordinates x, y, z has position described by the vector $\overrightarrow{0P}$, written $(x, y, z) \in \mathbf{R}^3$, or $x\boldsymbol{i} + y\boldsymbol{j} + z\boldsymbol{k}$, as convenient.

Direction of a Line. Let the line L make angles of α, β, γ, with the axes. This information together with one point lieing on L is sufficient to determine L. We call $\cos\alpha$, $\cos\beta$, $\cos\gamma$ the direction cosines of the line L.

A unit vector $\boldsymbol{l}$ in the direction of L can be written:

$$\boldsymbol{l} = (\cos\alpha)\,\boldsymbol{i} + (\cos\beta)\,\boldsymbol{j} + (\cos\gamma)\,\boldsymbol{k}.$$

Thus $\quad 1 = \boldsymbol{l}\cdot\boldsymbol{l} = \cos^2\alpha + \cos^2\beta + \cos^2\gamma.$

Sometimes we only know the proportional direction cosines (direction numbers or direction ratios) of L.

Thus if $n_1 : n_2 : n_3$ are direction numbers of L, then $\boldsymbol{n} = n_1\boldsymbol{i} + n_2\boldsymbol{j} + n_3\boldsymbol{k}$ is a vector in the direction of the line but is not necessarily a unit vector.

$$\boldsymbol{n} \cdot \boldsymbol{n} = n_1^2 + n_2^2 + n_3^2, \quad \frac{1}{|\boldsymbol{n}|}\, \boldsymbol{n} = (\cos\alpha)\,\boldsymbol{i} + (\cos\beta)\,\boldsymbol{j} + (\cos\gamma)\,\boldsymbol{k}.$$

$$|\boldsymbol{n}| = \sqrt{n_1^2 + n_2^2 + n_3^2} \text{ is the length of } \boldsymbol{n}.$$

Thus $\quad \cos\alpha = n_1/|\boldsymbol{n}|, \quad \cos\beta = n_2/|\boldsymbol{n}|, \quad \cos\gamma = n_3/|\boldsymbol{n}|.$

5.2 EQUATION OF A LINE

Let $\overrightarrow{OP} = \boldsymbol{p}$. Let $\boldsymbol{a}$ be any non-zero vector in the direction of the line. Then any point on the line through P in the direction of $\boldsymbol{a}$ is described by the position vector $\boldsymbol{r} = \boldsymbol{p} + \lambda\boldsymbol{a}$, for arbitrary $\lambda \in \mathbf{R}$.

In terms of coordinates we have:

$$\boldsymbol{r} = x\boldsymbol{i} + y\boldsymbol{j} + z\boldsymbol{k}, \quad \boldsymbol{p} = p_1\boldsymbol{i} + p_2\boldsymbol{j} + p_3\boldsymbol{k}, \quad \boldsymbol{a} = a_1\boldsymbol{i} + a_2\boldsymbol{j} + a_3\boldsymbol{k}.$$

Thus $\quad x = p_1 + \lambda a_1, \; y = p_2 + \lambda a_2, \; z = p_3 + \lambda a_3.$

Thus equation of line is:

$$\boldsymbol{r} = \boldsymbol{p} + \lambda\boldsymbol{a}, \text{ or}$$

$$\frac{x - p_1}{a_1} = \frac{y - p_2}{a_2} = \frac{z - p_3}{a_3}.$$

$a_1 : a_2 : a_3$ are the direction ratios of the line. The latter equation is often called the symmetric equation of the line.

Figure 5.2.1 illustrates the situation.

Alternatively, a line may be specified by giving two points lieing on it. Let P and Q be two points on the line L, where $\overrightarrow{OP} = \boldsymbol{p}$ and $\overrightarrow{OQ} = \boldsymbol{q}$.

Then $\overrightarrow{PQ}$ is a vector in the direction of the line.

Now $\overrightarrow{PQ} = \boldsymbol{q} - \boldsymbol{p}$. Thus $\boldsymbol{r} = \boldsymbol{p} + \lambda(\boldsymbol{q} - \boldsymbol{p})$ gives the line.

EXAMPLE 5.2.1 Find the line through the points (4,1,–3) and (2,0,1).

Solution. $r = (2,0,1) + \lambda(2,1,-4)$, or $\begin{cases} x = 2 + \lambda 2 \\ y = 0 + \lambda \\ z = 1 + \lambda(-4) \end{cases}$

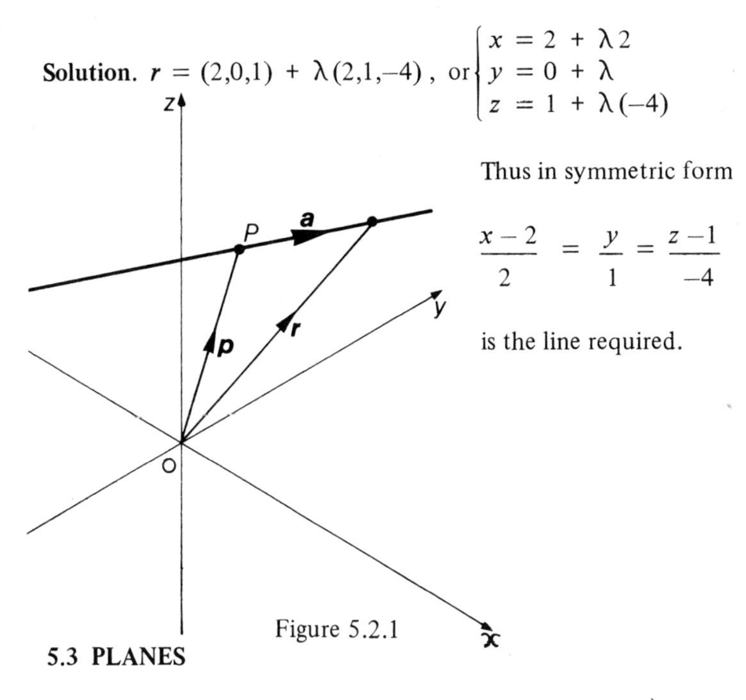

Figure 5.2.1

Thus in symmetric form

$$\frac{x-2}{2} = \frac{y}{1} = \frac{z-1}{-4}$$

is the line required.

5.3 PLANES

Let P be some point on the plane. Let the vector $\overrightarrow{OP} = \boldsymbol{p}$. Take any two directions $\boldsymbol{a}$ and $\boldsymbol{b}$ in the plane. Let $\boldsymbol{r}$ be the position vector of any point in the plane. Then $\boldsymbol{r} = \boldsymbol{p} + \lambda \boldsymbol{a} + \mu \boldsymbol{b}$, where λ and μ are arbitrary real numbers.

If P, Q, and T are non-collinear points, with $\overrightarrow{OP} = p$, $\overrightarrow{OQ} = q$ and $\overrightarrow{OT} = t$, then $\boldsymbol{q} - \boldsymbol{p}$ and $\boldsymbol{t} - \boldsymbol{p}$ may be taken as $\boldsymbol{a}$ and $\boldsymbol{b}$ above. The plane through P, Q, and T is then:

$$\boldsymbol{r} = \boldsymbol{p} + \lambda(\boldsymbol{q} - \boldsymbol{p}) + \mu(\boldsymbol{t} - \boldsymbol{p}).$$

Alternatively, let $\boldsymbol{n}$ be any vector perpendicular to the plane. Then $\boldsymbol{n} \cdot \boldsymbol{a} = 0$ and $\boldsymbol{n} \cdot \boldsymbol{b} = 0$, where here we use the scalar product or dot product of two geometric vectors.

Thus $\boldsymbol{n} \cdot \boldsymbol{r} = \boldsymbol{n} \cdot \boldsymbol{p}$, which in coordinate form is $n_1 x + n_2 y + n_2 z = d$. The d is short for $\boldsymbol{n} \cdot \boldsymbol{p}$.

Conversely any equation of the form $\boldsymbol{n} \cdot \boldsymbol{r} = \boldsymbol{n} \cdot \boldsymbol{p}$ may be written $\boldsymbol{n} \cdot (\boldsymbol{r} - \boldsymbol{p}) = 0$. Thus $(\boldsymbol{r} - \boldsymbol{p})$ is perpendicular to $\boldsymbol{n}$ and therefore lies in the plane through $\boldsymbol{p}$ perpendicular to $\boldsymbol{n}$.

Hence $\boldsymbol{n} \cdot \boldsymbol{r} = \boldsymbol{n} \cdot \boldsymbol{p}$ represents a plane through $\boldsymbol{p}$ perpendicular to $\boldsymbol{n}$.

If $\boldsymbol{n}$ is a unit vector, $\boldsymbol{p} \cdot \boldsymbol{n}$ is the distance from the origin to the plane.

Thus $n_1 x + n_2 y + n_3 z = d$ is the equation of a plane with $n_1 : n_2 : n_3$ the direction ratios of the normal to the plane. If n_1, n_2, n_3 are the actual direction cosines of the normal then d is the distance from the origin to the plane.

Note that if $\boldsymbol{a}$ and $\boldsymbol{b}$ lie in the plane, then the vector product $\boldsymbol{a} \wedge \boldsymbol{b}$ is perpendicular to the plane. This is a very useful result for finding normals to planes and hence the planes themselves when given suitable initial conditions. This will become apparent in the exercises.

Planes through Line of Intersection of Two Planes.
It is often useful to be able to write down the equation of any plane which has a common line of intersection with two other planes.

Let the given intersecting planes be

$$a_1 x + b_1 y + c_1 z = d_1$$

$$a_2 x + b_2 y + c_2 z = d_2$$

Consider $\quad (a_1 x + b_1 y + c_1 z - d_1) + \lambda(a_2 x + b_2 y + c_2 z - d_2) = 0.$

This is a plane and is satisfied by all points lieing on both planes. Suppose M is any plane through the line L defined by the two given planes. Let $P=(u, v, w)$ be any point on M **not** on the plane $a_2 x + b_2 y + c_2 z - d_2 = 0$.

Then $\quad (a_1 u + b_1 v + c_1 w - d_1) + \lambda(a_2 u + b_2 v + c_2 w - d_2) = 0$

determines λ. But then this value of λ gives the plane M. Thus by varying λ **we** get all the planes through L except for $a_2 x + b_2 y + c_2 z = d_2$ itself which requires λ = infinity.

Perpendicular Distance from Point to Plane.

Let plane be $\quad ax + by + cz = d.$

Let point be $\quad P = (x_1, y_1, z_1).$

Then the line $\quad r = (x_1, y_1, z_1) + \lambda(a, b, c)$ cuts the plane

when $\quad a(x_1 + \lambda a) + b(y_1 + \lambda b) + c(z_1 + \lambda c) = d.$

This determines λ to be $-((ax_1 + by_1 + cz_1) + d) / (a^2 + b^2 + c^2)$. Now the foot Q of the perpendicular from P onto the plane is:

$$(x_1 + \lambda a,\ y_1 + \lambda b,\ z_1 + \lambda c).$$

Thus the distance PQ is:

$$\left| \sqrt{(\lambda a)^2 + (\lambda b)^2 + (\lambda c)^2} \right| = \left| \lambda\sqrt{a^2 + b^2 + c^2} \right| =$$

$$= \left| (ax_1 + by_1 + cz_1 - d) \Big/ \sqrt{a^2 + b^2 + c^2} \right|$$

5.4 LINES AGAIN

First we find the common perpendicular to two non-intersecting lines.

Let L and M be the two lines with equations:

$$\boldsymbol{r} = \boldsymbol{p} + \lambda \boldsymbol{a} \quad (\boldsymbol{a} \neq \boldsymbol{0}) \text{ and } \boldsymbol{r} = \boldsymbol{q} + \mu \boldsymbol{b} \quad (\boldsymbol{b} \neq \boldsymbol{0}).$$

If P is the point $\boldsymbol{p} + \lambda\, \boldsymbol{q}$ on L and Q is the point $\boldsymbol{q} + \mu\, \boldsymbol{b}$ on M, then the line segment $\overrightarrow{PQ}$ is

$$(\boldsymbol{q} + \mu \boldsymbol{b}) - (\boldsymbol{p} + \lambda \boldsymbol{a}) = (\boldsymbol{q} - \boldsymbol{p} + \mu \boldsymbol{b} - \lambda \boldsymbol{a}).$$

The condition that $\overrightarrow{PQ}$ is perpendicular to both L and M is:

$$(\boldsymbol{q} - \boldsymbol{p} + \mu \boldsymbol{b} - \lambda \boldsymbol{a}) \cdot \boldsymbol{a} = 0 \quad \text{and}$$

$$(\boldsymbol{q} - \boldsymbol{p} + \mu \boldsymbol{b} - \lambda \boldsymbol{a}) \cdot \boldsymbol{b} = 0.$$

Thus $\quad \lambda(\boldsymbol{a} \cdot \boldsymbol{a}) - \mu(\boldsymbol{b} \cdot \boldsymbol{a}) = (\boldsymbol{q} \cdot \boldsymbol{a}) - (\boldsymbol{p} \cdot \boldsymbol{a})$

and $\quad \lambda(\boldsymbol{a} \cdot \boldsymbol{b}) - \mu(\boldsymbol{b} \cdot \boldsymbol{b}) = (\boldsymbol{q} \cdot \boldsymbol{b}) - (\boldsymbol{p} \cdot \boldsymbol{b}).$

These are linear equations for λ and μ.

If $D = \begin{vmatrix} \boldsymbol{a} \cdot \boldsymbol{a} & -\boldsymbol{b} \cdot \boldsymbol{a} \\ \boldsymbol{a} \cdot \boldsymbol{b} & -\boldsymbol{b} \cdot \boldsymbol{b} \end{vmatrix} = -(\boldsymbol{a} \cdot \boldsymbol{a})(\boldsymbol{b} \cdot \boldsymbol{b}) + (\boldsymbol{a} \cdot \boldsymbol{b})^2$

$$= -|\boldsymbol{a}|^2\, |\boldsymbol{b}|^2 + |\boldsymbol{a}|^2 |\boldsymbol{b}|^2 \cos^2 \theta, \text{ where } \boldsymbol{a} \cdot \boldsymbol{b} = |\boldsymbol{a}|\, |\boldsymbol{b}| \cos \theta,$$

is non-zero, then there is a unique solution.

From the above $D = (\cos^2 \theta - 1)\,|\boldsymbol{a}|^2\,|\boldsymbol{b}|^2$, which is non-zero provided $\cos \theta \neq 1$. Thus there is a unique common perpendicular when the lines are not parallel.

When the lines are parallel we have $\boldsymbol{a} = k\boldsymbol{b}$, k scalar, and

$$\lambda k^2 (\boldsymbol{b} \cdot \boldsymbol{b}) - \mu k (\boldsymbol{b} \cdot \boldsymbol{b}) = k(\boldsymbol{q} \cdot \boldsymbol{b}) - k(\boldsymbol{p} \cdot \boldsymbol{b}).$$

Thus $\lambda k\,(b \cdot b) - \mu\,(b \cdot b) = (q \cdot b) - (p \cdot b)$

is the only equation. Hence there are an infinity of solutions as expected.

Solving the equations above gives λ and μ for the points P and Q. The equation of the line is then just that for a line joining P and Q.

Length of the Common Perpendicular.
This is, of course, the distance PQ. If, however, only this length is required, it is not necessary to find the equation of PQ. Take any C on the line L and any D on the line M.

If $\boldsymbol{l}$ is the direction of L and $\boldsymbol{m}$ is the direction of M, then $(\boldsymbol{l} \wedge \boldsymbol{m})/|\boldsymbol{l} \wedge \boldsymbol{m}| = \boldsymbol{n}$ is a unit vector perpendicular to both lines.

Thus length of PQ is $\boldsymbol{n} \cdot \overrightarrow{CD}$.

Shortest Distance.
If the lines L and M do not intersect, let A and B be two planes with the common perpendicular as normal and with L in A and M in B. The planes are parallel. The common perpendicular gives the least distance between the planes. Hence this must also be the shortest distance between the two lines.

Note that the angle θ between two intersecting lines with directions given by the vectors $\boldsymbol{l}$ and $\boldsymbol{m}$ may be calculated from $\cos\theta = (\boldsymbol{l} \cdot \boldsymbol{m})/|\boldsymbol{l}|\,|\boldsymbol{m}|$.

If the lines are oriented we take $0^\circ \leqq \theta \leqq 180^\circ$, otherwise we take $0^\circ \leqq \theta \leqq 90^\circ$, where angles are in degrees.

Division of a Line.
We want to find a point on a segment of a line dividing it in a given ratio. Let the given segment be $\overrightarrow{PQ}$. Let $\overrightarrow{OP} = \boldsymbol{p}$ and $\overrightarrow{OQ} = \boldsymbol{q}$.

We want T where $|PT| : |TQ| = \lambda : 1$.

$$\overrightarrow{PT} = \frac{\lambda}{1+\lambda}\,\overrightarrow{PQ} = \frac{\lambda}{\lambda+1}\,(\boldsymbol{q}-\boldsymbol{p}).$$

Suppose that P has coordinates (x_1, y_1, z_1) and Q has coordinates (x_2, y_2, z_2).

Then $\overrightarrow{PQ} = (x_2 - x_1)\boldsymbol{i} + (y_2 - y_1)\boldsymbol{j} + (z_2 - z_1)\boldsymbol{k}$.

Thus $\overrightarrow{PT} = \dfrac{\lambda}{\lambda+1}\,\overrightarrow{PQ}$.

Hence if T is (x, y, z), we have:

$$(x - x_1)\boldsymbol{i} + (y - y_1)\boldsymbol{j} + (z - z_1)\boldsymbol{k} = \frac{\lambda}{\lambda+1}\{(x_2 - x_1)\boldsymbol{i} + (y_2 - y_1)\boldsymbol{j} + (z_2 - z_1)\boldsymbol{k}\}.$$

Thus $x = (\lambda x_2 + x_1)/(\lambda + 1)$

$y = (\lambda y_2 + y_1)/(\lambda + 1)$

$z = (\lambda z_2 + z_1)/(\lambda + 1)$.

5.5 SPHERES

Let sphere have centre given by the position vector $\boldsymbol{a}$, and radius given by the scalar k. Then the equation of the sphere is:

$$(\boldsymbol{r} - \boldsymbol{a})^2 = k^2, \text{ or } |\boldsymbol{r} - \boldsymbol{a}| = k,$$

where $\boldsymbol{r}$ is the position vector of an arbitrary point on the surface of the sphere.

If $\boldsymbol{r} = (x, y, z) = x\boldsymbol{i} + y\boldsymbol{j} + z\boldsymbol{k}$ and $\boldsymbol{a} = a_1\boldsymbol{i} + a_2\boldsymbol{j} + a_3\boldsymbol{k}$

then the equation is:

$$(x - a_1)^2 + (y - a_2)^2 + (z - a_3)^2 = k^2.$$

Tangent Plane to a Sphere.
Let the point on the sphere be P. Let $\vec{OP} = \boldsymbol{b}$. Let sphere have centre $\boldsymbol{a}$. Then $\boldsymbol{r}$ will be the position vector of a point on the tangent plane to the sphere at P if and only if $\boldsymbol{a} - \boldsymbol{b}$ is perpendicular to $\boldsymbol{r} - \boldsymbol{b}$. Thus the equation of the tangent plane is:

$$(\boldsymbol{r} - \boldsymbol{b}) \cdot (\boldsymbol{a} - \boldsymbol{b}) = 0.$$

Figure 5.5.1 illustrates the situation.

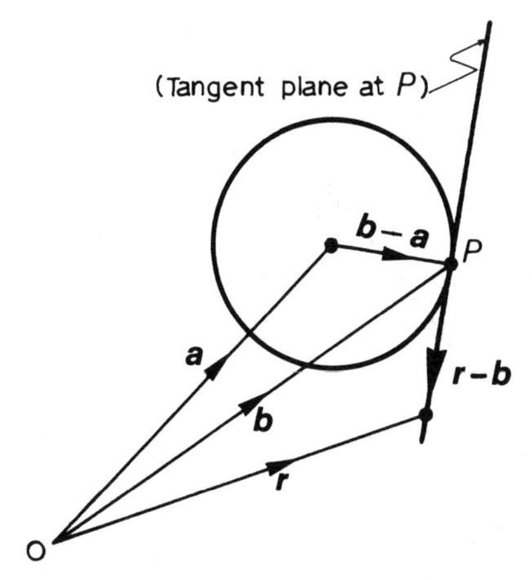

Figure 5.5.1

5.6 COORDINATE TRANSFORMATIONS

All systems of coordinates are Cartesian as described in section 5.1. We shall be concerned with two basic transformations from one Cartesian system to another.

(1) A shift of the origin while the new axes are parallel to the old. This is usually called a **translation**.

(2) The origin remains fixed. We usually require both sets of axes to be right handed. When this is so the change from one set of axes to the other may be looked on as a rotation about a line through the origin, as we shall see later.

Translation.
Let $Ox_1\,x_2\,x_3$ be the old system and $O'x_1{}'x_2{}'x_3{}'$ be the new system. Let the translation of the origin be given by the vector $\boldsymbol{a}$ relative to $Ox_1\,x_2\,x_3$. Then the equations describing the transformation are:

$$x_1{}' = x_1 - a_1$$

$$x_2{}' = x_2 - a_2$$

$$x_3{}' = x_3 - a_3$$

where $\boldsymbol{a} = a_1\boldsymbol{i} + a_2\boldsymbol{j} + a_3\boldsymbol{k}$ relative to the $Ox_1\,x_2\,x_3$ system.

Figure 5.6.1 illustrates the situation.

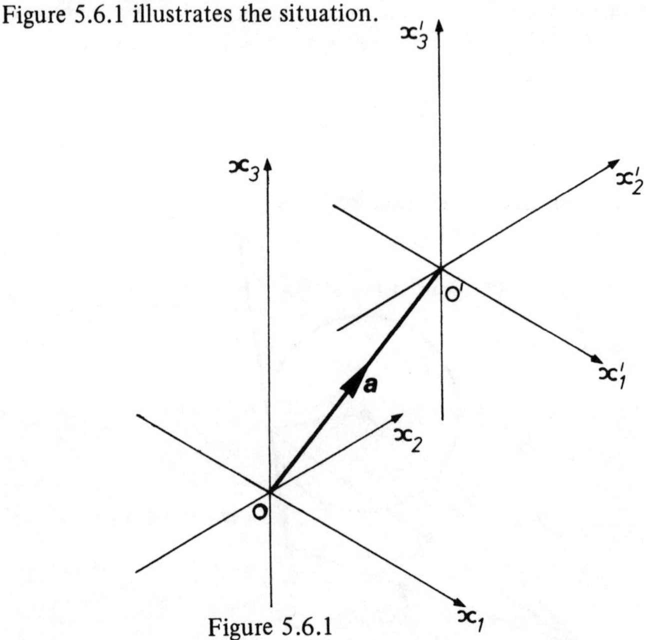

Figure 5.6.1

Fixed Origin.
Now suppose we have two sets of axes with the same origin O. Let the two systems be $Ox_1\,x_2\,x_3$ and $Ox_1{}'x_2{}'x_3{}'$. Let a_{rs} be the direction cosine of $x_r{}'$ with respect to x_s. Let $\theta_{rs} = \cos^{-1} a_{rs}$. Here $r = 1,2,3$ and $s = 1,2,3$.

Let P be a point in space. We describe $\vec{OP}$ relative to the two sets of axes in order to find the equations connecting the two sets of axes.

Figure 5.6.2 may help to visualise the situation.

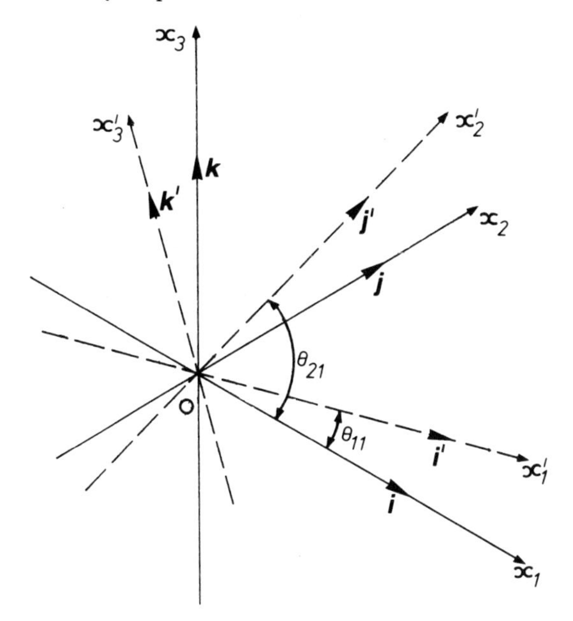

Figure 5.6.2

$$\vec{OP} = x_1 i + x_2 j + x_3 k = x_1' i' + x_2' j' + x_3' k' \tag{1}$$

$$\begin{aligned} i &= a_{11} i' + a_{21} j' + a_{31} k' \\ j &= a_{12} i' + a_{22} j' + a_{32} k' \\ k &= a_{13} i' + a_{23} j' + a_{33} k' \end{aligned} \quad \text{and} \quad \begin{aligned} i' &= a_{11} i + a_{12} j + a_{13} k. \\ j' &= a_{21} i + a_{22} j + a_{23} k. \\ k' &= a_{31} i + a_{32} j + a_{33} k. \end{aligned} \tag{2}$$

Using (1) we get by substitution from (2):

$$x_1 (a_{11} i' + a_{21} j' + a_{31} k') + x_2 (a_{12} i' + a_{22} j' + a_{32} k') + x_3 (a_{13} i' + a_{23} j' + a_{33} k') =$$

$$= x_1' i' + x_2' j' + x_3' k'.$$

Comparing coefficients we have:

$$\begin{aligned} x_1' &= a_{11} x_1 + a_{12} x_2 + a_{13} x_3 \\ x_2' &= a_{21} x_1 + a_{22} x_2 + a_{23} x_3 \\ x_3' &= a_{31} x_1 + a_{32} x_2 + a_{33} x_3 \end{aligned} \tag{3}$$

Thus $\quad x' = A\,x$, where $x' = \begin{pmatrix} x_1' \\ x_2' \\ x_3' \end{pmatrix}, \quad x = \begin{pmatrix} x_1 \\ x_2 \\ x_3 \end{pmatrix}$ and

$$A = \begin{pmatrix} a_{11} & a_{12} & a_{13} \\ a_{21} & a_{22} & a_{23} \\ a_{31} & a_{32} & a_{33} \end{pmatrix} .$$

Now if F is any field and F^n and F^n_c are as defined in chapter 4, then we say that two vectors $x, y \in F^n$ are **orthogonal** if $x_1 y_1 + x_2 y_2 + \ldots\ldots + x_n y_n = 0$, where $x = (x_1, x_2, \ldots\ldots, x_n)$ and $y = (y_1, y_2, \ldots\ldots, y_n)$. We say that x has **unit length** if $x_1^2 + x_2^2 + \ldots\ldots + x_n^2 = 1$.

Similar definitions hold for vectors in F^n_c.

From $\quad i^2 = j^2 = k^2 = 1$ and $i \cdot j = i \cdot k = j \cdot k = 0$

and $\quad i'^2 = j'^2 = k'^2 = 1$ and $i' \cdot j' = i' \cdot k' = j' \cdot k' = 0,$

we deduce that the columns of A are mutually orthogonal as vectors in $\mathbf{R}^3_c$ and have unit length and that the rows of A are mutually othogonal as vectors in $\mathbf{R}^3$ and have unit length. For example:

$$i^2 = (a_{11}i' + a_{21}j' + a_{31}k') \cdot (a_{11}i' + a_{21}j' + a_{31}k').$$

Thus $\quad 1 = a_{11}^2 + a_{21}^2 + a_{31}^2.$

Hence column $\begin{pmatrix} a_{11} \\ a_{21} \\ a_{31} \end{pmatrix}$ of A has unit length.

Orthogonal Matrix.

A set of unit vectors which are mutually orthogonal is called an **orthonormal** set and the vectors are said to be **orthonormal** vectors.

Let A be an $n \times n$ matrix over the field F. A is said to be an **orthogonal** matrix if **any one** of the following equivalent conditions holds.

(1) The columns of A are orthonormal.

(2) The rows of A are orthonormal.

(3) $A' = A^{-1}$.

(4) $A'A = I_n$.

(5) $AA' = I_n$.

The equivalence of these conditions is easy to see if the reader writes out A in some detail and thinks for a moment what each condition is saying.

In particular

$$A = \begin{pmatrix} a_{11} & a_{12} & a_{13} \\ a_{21} & a_{22} & a_{23} \\ a_{31} & a_{32} & a_{33} \end{pmatrix}$$

associated with the change of coordinates with fixed origin is an **orthogonal** matrix.

The determinant of an orthogonal matrix has a particularly simple value. Condition (4) above says that for such a matrix $A' A = I_n$.

Thus $|I_n| = |A'A| = |A'||A| = |A|^2$.

Hence $|A|^2 = 1$.

Thus $|A| = \pm 1$.

Rotations.

Note that we can get from the coordinate system $Ox_1 \, x_2 \, x_3$ to the coordinate system $Ox_1{}' x_2{}' x_3{}'$ in at most three steps as follows.

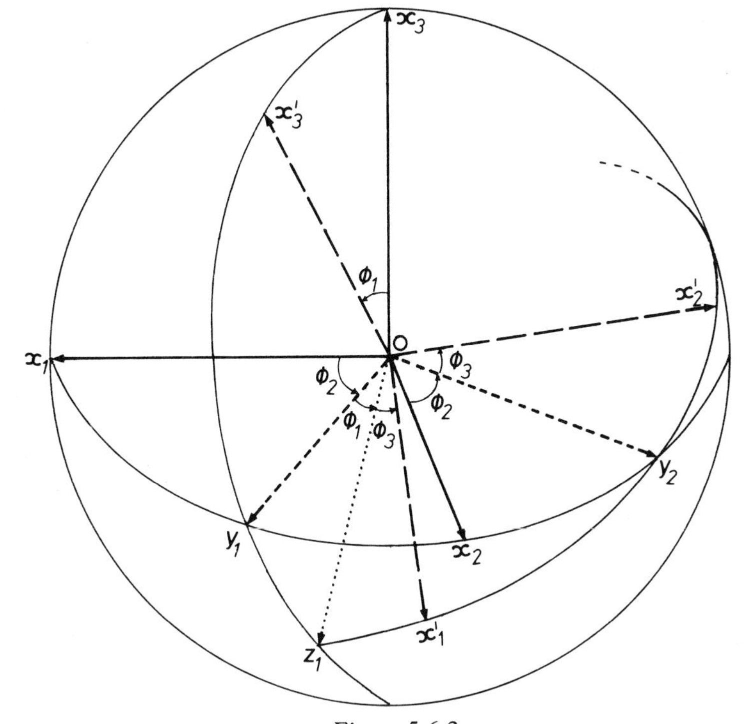

Figure 5.6.3

Figure 5.6.3 shows the two coordinate systems drawn in a sphere in order to make clear the various rotations required.

Let ϕ_1 be the angle between Ox_3 and $Ox_3{}'$. Let ϕ_2 be the angle between the plane $Ox_3\,x_3{}'$ and the plane $Ox_1\,x_3$. Finally let ϕ_3 be the angle between the planes $Ox_1{}'\,x_3{}'$ and $Ox_3\,x_3{}'$.

(1) Rotate $Ox_1\,x_2\,x_3$ through ϕ_2 about the axis Ox_3. This takes the coordinate system $Ox_1\,x_2\,x_3$ into the system $Oy_1\,y_2\,x_3$.

(2) Rotate the system $Oy_1\,y_2\,x_3$ through an angle ϕ_1 about the axis Oy_2. This takes us into the system $Oz_1\,y_2\,x_3{}'$.

(3) Rotate through ϕ_3 about the axis $Ox_3{}'$. This takes us to the coordinate system $Ox_1{}'\,x_2{}'\,x_3{}'$ as required.

At each stage we are changing coordinates by a rotation about some coordinate axis.

Let us look more closely at the matrix associated with this kind of transformation. Figures 5.6.4 and 5.6.5 may help to see what is involved.

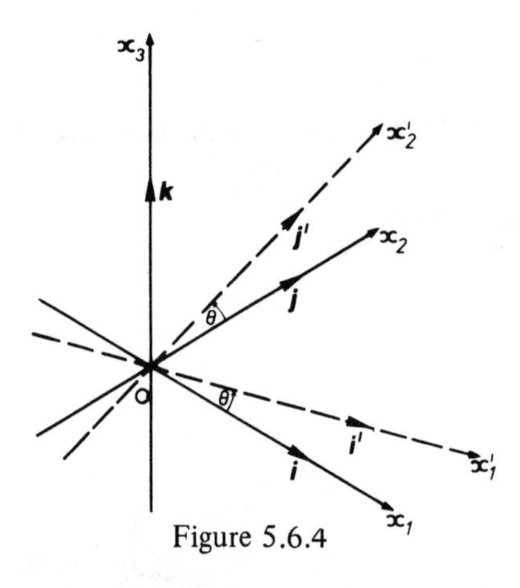

Figure 5.6.4

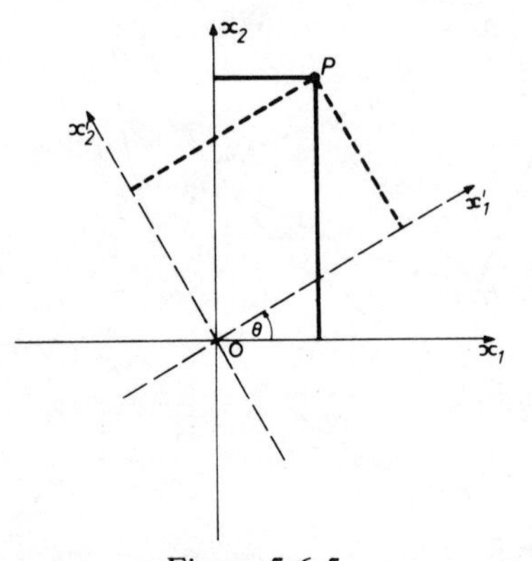

Figure 5.6.5

Let us rotate about the x_3-axis through an angle θ in the anti-clockwise sense. Then we have the transformation:

$$x_1{}' = a_{11}\,x_1 + a_{12}\,x_2$$

$$x_2{}' = a_{21}\,x_1 + a_{22}\,x_2$$

By our previous result for the general transformation, this gives:

$$a_{11} = \cos\theta, \quad a_{21} = \cos(\theta + {}^{\pi}/_{2})$$

$$a_{12} = \cos({}^{\pi}/_{2} - \theta), \quad a_{22} = \cos\theta$$

since $a_{rs} = \cos\theta_{rs}$.

Thus $\boldsymbol{x}' = \begin{pmatrix} \cos\theta & \sin\theta \\ -\sin\theta & \cos\theta \end{pmatrix} \boldsymbol{x}$;

$$\boldsymbol{x}' = \begin{pmatrix} x_1{}' \\ x_2{}' \end{pmatrix}, \quad \boldsymbol{x} = \begin{pmatrix} x_1 \\ x_2 \end{pmatrix}.$$

In three dimensions this takes the form:

$$x' = \begin{pmatrix} \cos\theta & \sin\theta & 0 \\ -\sin\theta & \cos\theta & 0 \\ 0 & 0 & 1 \end{pmatrix} x = A_1 x, \text{ and } |A_1| = 1.$$

The general transformation considered earlier has the matrix $A = A_1 A_2 A_3$, since it can be obtained by a sequence of at most three transformations of the special kind considered above. Since $|A_1| = |A_2| = |A_3| = 1$ and $|A| = |A_1|\,|A_2|\,|A_3|$, it follows that $|A| = 1$.

We conclude that the general transformation from a right hand system to a right hand system is described by $x' = A x$ with A orthogonal and $|A| = 1$.

Transformation Given by $x' = A x$ with A orthogonal of Determinant 1.

We now show the converse.

Suppose we change our coordinates according to the equations $x' = A x$ with A orthogonal and $|A| = 1$. Can we interpret the new coordinates as a right hand Cartesian system?

Since A is orthogonal we can interpret a_{rs} as the cosine of an angle θ_{rs}. Taking coordinate lines x_1', x_2', x_3' making the appropriate angles with Ox_1, Ox_2 and Ox_3 we get a new coordinate system $Ox_1'x_2'x_3'$. Taking unit vectors i', j', k' along these axes we get:

$$\begin{aligned} i' &= a_{11}i + a_{12}j + a_{13}k \\ j' &= a_{21}i + a_{22}j + a_{23}k \\ k' &= a_{31}i + a_{32}j + a_{33}k \end{aligned}$$

Then $\quad i'^2 = j'^2 = k'^2 = 1$ and $i' \cdot j' = i' \cdot k' = j' \cdot k' = 0.$

Thus our new axes are mutually orthogonal.

If the new axes form a left hand system, then changing the direction of one axis, say Ox_3' will give a right hand system. Now this involves changing the sign of one row of A. Thus the matrix A_1 associated with our new right hand system will have $|A_1| = -1$. However, this cannot be, because the matrix of the transformation from a right hand system to a right hand system must have determinant 1, as we have seen. It follows that $Ox_1'x_2'x_3'$ defined above is a right hand system of mutually orthogonal axes.

Summary.

The foregoing considerations show that the transformation from one right hand system to another is described by $x' = A x$, with A orthogonal and $|A| = 1$. Conversely, if $Ox_1 x_2 x_3$ is a right hand system of mutually orthogonal axes, and we are given a change of coordinates according to the equations $x' = Ax$, then the new system of coordinates $Ox_1'x_2'x_3'$ can be taken to form a right hand system of mutually orthogonal axes.

Note that if A is an orthogonal matrix with $|A| = 1$, then the transformation $\boldsymbol{x}' = A\,\boldsymbol{x}$ can be thought of as a rotation of the triad of axes about some line through the origin O.

This follows from the fact that the transformation A, considered as a linear transformation of $\mathbf{R}^3$, leaves distances, or lengths of vectors, unchanged. To see this, consider the effect of A on the vector $\boldsymbol{x}$. In fact $\boldsymbol{x}$ becomes $A\,\boldsymbol{x}$. Now the length of

$$\boldsymbol{x} = \begin{pmatrix} x_1 \\ x_2 \\ x_3 \end{pmatrix} \quad \text{is} \sqrt{(x_1^2 + x_2^2 + x_3^2)} = \sqrt{(\boldsymbol{x}'\,\boldsymbol{x})}$$

and the length of $A\,\boldsymbol{x}$ is

$$\sqrt{(A\boldsymbol{x})'\,(A\boldsymbol{x})},$$

where the products are matrix products (in this context sometimes called inner products of the vectors involved). Now we have:

$$(A\,\boldsymbol{x})'\,(A\,\boldsymbol{x}) = \boldsymbol{x}'A'A\,\boldsymbol{x} = \boldsymbol{x}'I_3\boldsymbol{x} = \boldsymbol{x}'\boldsymbol{x},$$

since A is orthogonal. Thus the length of $\boldsymbol{x}$ is unchanged after transformation by A.

Also $\quad |A - I_3| = |A' - I_3| = |A^{-1} - I_3| = |A^{-1}(I_3 - A)| =$

$$= |A^{-1}|\,|I_3 - A| = |A^{-1}|\,|-I_3(A - I_3)| = |A'|\,|-I_3|\,|A - I_3| =$$

$$= 1 \cdot (-1)^3 \cdot |A - I_3| = -|A - I_3|.$$

Hence $\quad 2|A - I_3| = 0.$

Thus $\quad |A - I_3| = 0.$

This means that the system of linear equations

$$(A - I_3)\begin{pmatrix} x_1 \\ x_2 \\ x_3 \end{pmatrix} = 0$$

has a non-zero solution. Thus there is a vector

$$\boldsymbol{v} = \begin{pmatrix} v_1 \\ v_2 \\ v_3 \end{pmatrix}, \quad \text{such that } A\,\boldsymbol{v} = \boldsymbol{v}.$$

Hence the line through the origin in the direction of $\boldsymbol{v}$ is fixed at every point by A.

Altogether we have shown that A fixes lengths or distances and also fixes every point in a line through the origin 0. Thus A must be a rotation around the fixed line $\boldsymbol{r} = \lambda \boldsymbol{v}$, λ arbitrary. In this case we can express the line as the subspace $\langle \boldsymbol{v} \rangle$ of $\mathbf{R}^3$, because this line contains the origin.

EXERCISES

1. Find equations for the lines through the pairs of points with the following coordinates:

(a) (1,1,1) and (4,2,3) (b) (2,0,1) and (7,4,2)

(c) (1,4,2) and (0,5,3).

2. Find the angle between each pair of lines in exercise (1) above.

3. Find equations of the lines described below.

(a) Line through (0,1,–1) in the direction 1:2:3.

(b) Line through (–1,0,0) in the direction 1:–1:1.

(c) Line through (1,1,1) in the direction of the normal to the plane $2x - y + z = 1$.

4. Find the points where the common perpendicular to each of the following pairs of lines meets these lines.

(a) $\boldsymbol{r} = (1,4,0) + \lambda(0,1,1)$ and $\boldsymbol{r} = (1,-1,4) + \mu(2,1,2)$.

(b) $\dfrac{x}{1} = \dfrac{(y-1)}{1} = \dfrac{(z-1)}{1}$ and $\dfrac{(x-1)}{1} = \dfrac{y}{2} = \dfrac{z}{(-3)}$.

Hence find the shortest distance between the lines in each case.
Write down the equations of the common perpendiculars.

5. The direction cosines of a moving straight line in two neighbouring positions are l,m,n and $l + \delta l$, $m + \delta m$, $n + \delta n$. Find an expression for the small angle $\delta\theta$ between the two positions of the line.

6. Two lines L and M have direction ratios $l{:}m{:}n$ and $l'{:}m'{:}n'$. Find the direction ratios of a line N which is perpendicular to both L and M.

7. Find the points at which the line joining the points (5,–3,1) and (3,2,7) cuts the coordinate planes.

Also find the ratio in which this line is divided by the x–z coordinate plane i.e. the plane $y = 0$.

8. If (x_1, y_1, z_1), (x_2, y_2, z_2), (x_3, y_3, z_3) are three **distinct** points, show that the equation of the plane containing these points may be written:

$$\begin{vmatrix} x - x_1 & y - y_1 & z - z_1 \\ x_2 - x_1 & y_2 - y_1 & z_2 - z_1 \\ x_3 - x_1 & y_3 - y_1 & z_3 - z_1 \end{vmatrix} = 0.$$

9. Find the equation of the line of intersection of the two planes:
$3x - 2y + z = 1$ and $5x + 4y - 6z = 2$.

10. Find the distance from the point (0,1,1) to the plane: $x + y + z = 1$.

11. Find the planes through the following sets of points:

(a) (1,2,3), (3,5,7), (3,–1,–3)

(b) (4,–5,8), (1,2,3), (2,9,–4)

(c) (0,12,2), (2,–2,14), (1,2,3).

12. Find the equations of the lines of intersection of pairs of planes found in exercise 11. Also investigate the nature of the intersection of all three planes found in exercise 11.

13. If L is the line

$$\frac{(x-1)}{3} = \frac{(y-2)}{4} = \frac{(z-5)}{6} \quad \text{and } P \text{ is the point } (-2,1,3),$$

find the equation of the plane which contains P and L and also that of the plane through L perpendicular to this plane.

14. Find the angle between the common line of the planes $x + y - z = 1$, $2x - 3y + z = 2$, and the line joining the points (3,–1,2), (4,0,–1).

Find also the equation of a line through the origin which is perpendicular to **both** the above lines.

15. The plane $4x + 7y + 4z + 81 = 0$ is rotated through a right angle about its line of intersection with the plane $5x + 3y + 10z = 25$.

Find the equation of the plane in its new position.

Find also the distance between the feet of the perpendiculars drawn from the origin onto the plane in its two positions.

16. Find the equation of the sphere with centre $(1,-2,3)$ and radius 2.

17. Find the centre and radius of the sphere

$$2(x^2 + y^2 + z^2) - 6x + 8y - 8z = 1.$$

18. Find the equations of the spheres through the points:

(a) $(1,-1,1)$, $(0,1,2)$, $(2,3,0)$, $(5,2,4)$

(b) $(3,3,4)$, $(-3,0,1)$, $(-1,3,-2)$, $(4,3,-1)$.

19. Find the equation of the tangent plane at $(1,-2,2)$ to the sphere

$$x^2 + y^2 + z^2 = 9.$$

20. Find the equations of the tangent planes to the sphere

$$x^2 + y^2 + z^2 + 2x - 4y + 6z - 7 = 0$$

which intersect in the line $6x - 3y - 23 = 0 = 3z + 2$.

21. Find the equations of the spheres that pass through the circle

$x^2 + z^2 - 2x + 2z = 2$, $y = 0$ and touch the plane $y - z = 7$.

22. A right handed Cartesian coordinate system is rotated through an angle of 120 degrees about the line through the origin with direction cosines $1/\sqrt{3}$: $1/\sqrt{3}$: $1/\sqrt{3}$. Find the matrix A associated with the transformation of co-ordinates that results.

Show that this matrix is orthogonal and $|A| = 1$.

Show also that $A^3 = I_3$. (Take the rotation anti-clockwise relative to the direction $1/\sqrt{3}\,(i + j + k)$).

23. A coordinate transformation is given by:

$$\begin{pmatrix} x' \\ y' \\ z' \end{pmatrix} = \begin{pmatrix} \frac{1}{4} & \frac{\sqrt{3}}{4} & \frac{\sqrt{3}}{2} \\ \frac{\sqrt{3}}{2} & -\frac{1}{2} & 0 \\ \frac{\sqrt{3}}{4} & \frac{3}{4} & -\frac{1}{2} \end{pmatrix} \begin{pmatrix} x \\ y \\ z \end{pmatrix}$$

Does this give a rotation of the coordinate system? If so, find the axis of rotation relative to the original $Oxyz$ coordinate system. Also find the angle of rotation.

CHAPTER 6

Quadrics and Quadratic Forms

6.1 INTRODUCTION

In two dimensions the circle, ellipse, parabola, and hyperbola are examples of curves whose equations relative to the usual Cartesian axes $Ox_1 x_2$ take the form:

$$a_{11} x_1^2 + 2a_{12} x_1 x_2 + a_{22} x_2^2 + b_1 x_1 + b_2 x_2 + c = 0 ;$$

$$a_{ij}, b_i, c \in \mathbf{R}.$$

Such an equation is said to be of the second degree and, for this reason, the conics referred to above are known as second degree curves.

By a shift of the origin and a suitable rotation of the axes, the general equation above can be made to take much simpler forms for the different conics. For example, for an ellipse the equation can be made to take the simple form:

$$\frac{x_1^2}{a^2} + \frac{x_2^2}{b^2} = 1.$$

These simple forms are called **standard** or **canonical** forms. By considering all the different standard forms that are possible, we can classify the conics. As is well known, the curves that can actually be drawn on paper are the parabola, the ellipse (with the circle as a special case), the hyperbola and a pair of straight lines.

The canonical forms then provide a convenient framework within which each conic may be studied in detail.

Equally important in three dimensions are the surfaces described by second degree equations relative to Cartesian axes. These surfaces are called **quadrics**.

Just as with the conics, we are interested in finding out what these surfaces look like. That is we want to classify them. As with the conics we look for canonical forms. It is clear that we can remove the linear terms, when possible, by a shift of the origin. This we will consider later. For the moment we concentrate on the reduction of that part of the equation:

$$a_{11}x_1^2 + a_{22}x_2^2 + a_{33}x_3^2 + 2a_{12}x_1x_2 + 2a_{13}x_1x_3 + 2a_{23}x_2x_3 +$$

$$+ b_1x_1 + b_2x_2 + b_3x_3 + c = 0$$

which does not involve linear or constant terms i.e.

$$a_{11}x_1^2 + a_{22}x_2^2 + a_{33}x_3^2 + 2a_{12}x_1x_2 + 2a_{13}x_1x_3 + 2a_{23}x_2x_3.$$

In the case of two dimensions we know that a suitable rotation of axes reduces this type of expression to a sum of squares form. Could it be that a rotation of axes would do the trick in three dimensions also.

By section 5.6 we know that rotating axes corresponds to a change of variable by an orthogonal matrix of determinant 1.

This means that we can restate our immediate problem as follows. By a change of variable $\boldsymbol{x}' = A\boldsymbol{x}$ with A orthogonal and $|A| = 1$, can we transform

$$a_{11}x_1^2 + a_{22}x_2^2 + a_{33}x_3^2 + 2a_{12}x_1x_2 + 2a_{13}x_1x_3 + 2a_{23}x_2x_3$$

into the form $\quad b_{11}x_1'^2 + b_{22}x_2'^2 + b_{33}x_3'^2$

In the subsequent work we shall usually refer to the latter as **'sum of squares'** form.

As usual in mathematics we express this in a more general context. First a definition:

DEFINITION 6.1.1 A quadratic form in n variables is any expression of the form:

$$Q(x_1, x_2, \ldots\ldots, x_n) = \sum_{i,j=1}^{n} a_{ij}x_ix_j, \text{ where } a_{ij} = a_{ji} \in \mathbf{R}.$$

The matrix $A = (a_{ij})$ is called the **matrix of the form**. It is symmetric i.e. $A' = A$.

Note that we can write

$$Q(x_1, x_2, \ldots\ldots, x_n) = \boldsymbol{x}'A\boldsymbol{x}, \ \boldsymbol{x} = \begin{pmatrix} x_1 \\ x_2 \\ \cdot \\ \cdot \\ x_n \end{pmatrix}.$$

Now suppose that we change the variables according to:

$$x = Ty, \text{ where } x = \begin{pmatrix} x_1 \\ \cdot \\ \cdot \\ \cdot \\ x_n \end{pmatrix}, \quad y = \begin{pmatrix} y_1 \\ y_2 \\ \cdot \\ \cdot \\ \cdot \\ \cdot \\ y_n \end{pmatrix}.$$

Then $\quad Q(x_1, x_2, \ldots\ldots, x_n) = y'\, T'A\, Ty.$

Now this takes the sum of squares form:

$$b_{11} y_1^2 + b_{22} y_2^2 + b_{33} y_3^2 + \ldots\ldots + b_{nn} y_n^2$$

if and only if

$$T'A\,T = \begin{pmatrix} b_{11} & 0 & 0 \ldots 0 \\ 0 & b_{22} & 0 \ldots 0 \\ 0 & 0 & b_{33} \ldots 0 \\ \cdot & \cdot & \cdot \quad \cdot \\ 0 & 0 & \ldots\ldots b_{nn} \end{pmatrix}, \text{ a diagonal matrix.}$$

Thus our problem can be framed in the following way. Given a symmetric matrix A, can we find an orthogonal matrix T with $|T| = 1$ so that $T'A\,T$ is a diagonal matrix. Since T is orthogonal we note that $T' = T^{-1}$.

Before trying to solve this problem let us consider an example of a quadratic form and its reduction to sum of squares form though not by a method satisfying our criteria.

EXAMPLE 6.1.1

Let $\quad Q(x_1, x_2, x_3) = x_1^2 + 3x_2^2 + 2x_1 x_2 + 3x_1 x_3.$

Write $\quad Q(x_1, x_2, x_3) = (x_1 + x_2 + 3x_3/2)^2 + 2x_2^2 - 9/4 x_3^2 - 3x_2 x_3.$

Now consider $\quad 2x_2^2 - 9/4 x_3^2 - 3x_2 x_3 = 2(x_2^2 - 9/8 x_3^2 - 3/2 x_2 x_3) =$

$$= 2\{(x_2 - 3/4 x_3)^2 - 9/16 x_3^2 - 9/8 x_3^2\} = 2\{(x_2 - 3/4 x_3)^2 - 27/16 x_3^2\}$$

Thus $\quad Q = (x_1 + x_2 + 3x_3/2)^2 + 2(x_2 - 3x_3/4)^2 - 27/8 x_3^2.$

Put $\quad \begin{cases} x_1' = x_1 + x_2 + 3x_3/2. \\ x_2' = x_2 - 3x_3/4 \\ x_3' = x_3 \end{cases}$ Then $\quad x' = Px$; where

$$P = \begin{pmatrix} 1 & 1 & 3/2 \\ 0 & 1 & -3/4 \\ 0 & 0 & 1 \end{pmatrix}.$$

In the new variables $Q = x_1'^2 + 2x_2'^2 - {}^{27}/_8 x_3'^2$.

We note that $|P| = 1$ but P is **not** orthogonal.
Since $|P|$ is non-zero, P has an inverse. A change of variables satisfying this condition is called a **non-singular transformation**. It means that we can express the old variables in terms of the new variables in a unique way.

This method will always work. When there are no square terms we first carry out the transformation

$$\left.\begin{aligned} x_1 &= y_1 + y_2 \\ x_2 &= y_1 - y_2 \end{aligned}\right\} \text{, say.}$$

For example, if we have $Q = x_1 x_2 + x_2 x_3 + x_1 x_3$, then the above transformation introduces square terms and we then carry on as before. We combine the transformations to obtain the required change of variables.

However this approach does not in general solve our problem.

6.2 EIGENVALUES AND EIGENVECTORS

We want to find an orthogonal matrix T so that $T'AT = T^{-1}AT$ is diagonal. Now this may remind the reader of our remarks in section 4.17.
If we let A define a linear transformation of a vector space and then choose a suitable basis of this space it is possible that relative to this new basis the matrix A is transformed into diagonal form as required. Such a method which emphasises the role played by the bases involved is rather abstract and more appropriate to a second course in linear algebra; see for example 'Topics in Algebra' by I.N. Herstein (Wiley 1975). Here we shall follow a more computational approach with the emphasis on the matrices and F_c^n, or rather $\mathbf{R}_c^n$ since mostly we shall restrict ourselves to the field of real numbers in this chapter.

The vectors $\boldsymbol{x} \in \mathbf{R}_c^n$ that are of special importance for us are those which satisfy the equation $A\boldsymbol{x} = \lambda \boldsymbol{x}$ for some scalar λ, where $\boldsymbol{x} \neq \boldsymbol{0}$.

DEFINITION 6.2.1 Let A be an $n \times n$ matrix over the field F. Any scalar λ satisfying an equation $A\boldsymbol{x} = \lambda \boldsymbol{x}$, where $\boldsymbol{x} \neq \boldsymbol{0}$, is called an **eigenvalue** of the matrix A. Any $\boldsymbol{x} \neq \boldsymbol{0}$ corresponding to λ is called an **eigenvector** of A belonging to λ.

Clearly if $\boldsymbol{x}$ is an eigenvector for λ, then so is $k\boldsymbol{x}$ for any scalar $k \neq 0$. Often we arrange the value of k so that the eigenvector satisfies

$$+\sqrt{\boldsymbol{x}'\boldsymbol{x}} = 1.$$

Written more fully this reads $+\sqrt{(x_1^2 + x_2^2 + \ldots\ldots + x_n^2)} = 1,$

$$\text{where } \boldsymbol{x} = \begin{pmatrix} x_1 \\ x_2 \\ \cdot \\ \cdot \\ \cdot \\ x_n \end{pmatrix}.$$

Often $+\sqrt{x'x}$ is called the **length** or the **norm** of the vector $\boldsymbol{x}$ and is denoted by $||\boldsymbol{x}||$.
Such an eigenvector is said to be **normalised.**

$$\text{Suppose we have:} \quad A\boldsymbol{x} = \lambda \boldsymbol{x}, \text{ with } \boldsymbol{x} = \begin{pmatrix} x_1 \\ \cdot \\ \cdot \\ \cdot \\ x_n \end{pmatrix}.$$

Then $(A - \lambda I_n)\boldsymbol{x} = \boldsymbol{0}$. Now this is a system of n linear homogeneous equations in n variables. Hence a non-zero $\boldsymbol{x}$ exists to satisfy the system if and only if $|A - \lambda I_n| = 0$, by theorem 4.16.3.

This equation is called the eigenvalue equation or **characteristic** equation of A. The roots of this equation give the possible λ for which $A\boldsymbol{x} = \lambda\boldsymbol{x}$ for some $\boldsymbol{x} \neq \boldsymbol{0}$.

Before we can solve our main problem we require some properties of eigenvalues and eigenvectors. These we establish in a sequence of theorems.

THEOREM 6.2.1

A and $T^{-1}AT$ have the **same** eigenvalues.

PROOF

$$|T^{-1}AT - \lambda I_n| = |T^{-1}(A - \lambda I_n)T| =$$

$$= |T^{-1}|\,|A - \lambda I_n|\,|T| = |A - \lambda I_n|\,|T^{-1}|\,|T| =$$

$$= |A - \lambda I_n|\,|T^{-1}T| = |A - \lambda I_n|\,|I_n| = |A - \lambda I_n|.$$

The theorem follows.

THEOREM 6.2.2 Any set of mutually orthogonal non-zero vectors $\boldsymbol{x}_1, \boldsymbol{x}_2, \ldots\ldots, \boldsymbol{x}_r \in F_C^n$ (for this theorem F can be any field) is linearly independent. A similar result holds for F^n.

PROOF

Recall that two vectors $\boldsymbol{x}$ and $\boldsymbol{y}$ are orthogonal if and only if $\boldsymbol{x}'\boldsymbol{y} = 0$.

Suppose that $a_1 \boldsymbol{x}_1 + a_2 \boldsymbol{x}_2 + \ldots\ldots + a_r \boldsymbol{x}_r = \boldsymbol{0}$.

Then $\boldsymbol{x}_i' (a_1 \boldsymbol{x}_1 + \ldots\ldots + a_r \boldsymbol{x}_r) = \boldsymbol{x}_i' \begin{pmatrix} 0 \\ \cdot \\ \cdot \\ 0 \end{pmatrix} = 0, i = 1,2,\ldots, r.$

Hence $a_1 \boldsymbol{x}_1{}' \boldsymbol{x}_1 + a_2 \boldsymbol{x}_i' \boldsymbol{x}_2 + \ldots\ldots + a_r \boldsymbol{x}_i' \boldsymbol{x}_r = 0.$

But $\boldsymbol{x}_i$ and $\boldsymbol{x}_j$ are given to be orthogonal. Hence $\boldsymbol{x}_i' \boldsymbol{x}_j = 0$, provided $i \neq j$.

Thus $a_i \boldsymbol{x}_i' \boldsymbol{x}_i = 0.$

Now $\boldsymbol{x}_i' \boldsymbol{x}_i \neq 0$ because $\boldsymbol{x}_i$ is given to be non-zero.

Thus $a_i = 0.$

This holds for $i = 1,2,3, \ldots, r$. Thus the only way we can have $a_1 \boldsymbol{x}_1 + a_2 \boldsymbol{x}_2 + \ldots\ldots + a_r \boldsymbol{x}_r = \boldsymbol{0}$ is by taking all the a_i to be zero. By definition $\boldsymbol{x}_1, \boldsymbol{x}_2, \ldots\ldots, \boldsymbol{x}_r$ are linearly independent vectors.

The theorem follows.

THEOREM 6.2.3 The eigenvalues of a real symmetric $n \times n$ matrix A are all real numbers.

PROOF

(1) $A\boldsymbol{x} = \lambda \boldsymbol{x}$. Let $\overline{A}$ denote the complex conjugate of A.

We have $\overline{A} = \begin{pmatrix} \overline{a}_{11} & \overline{a}_{12} \ldots\ldots & \overline{a}_{1n} \\ \cdot & \cdot \ldots\ldots & \cdot \\ \cdot & \cdot \ldots\ldots & \cdot \\ \cdot & \cdot \ldots\ldots & \cdot \\ \overline{a}_{n1} & \overline{a}_{n2} \ldots\ldots & \overline{a}_{nn} \end{pmatrix}$

Similarly $\overline{\boldsymbol{x}} = \begin{pmatrix} \overline{x}_1 \\ \cdot \\ \cdot \\ \cdot \\ \overline{x}_n \end{pmatrix}$.

From (1) we get $A\bar{x} = \bar{\lambda}\bar{x}$ since $\bar{A} = A$. (A is real).

Transpose this latter equation and use the fact that A is symmetric hence $A' = A$. The result is:

(2) $\bar{x}'A = \bar{\lambda}\bar{x}'$.

From (1) $\bar{x}'Ax = \lambda\bar{x}'x$ and from (2) $\bar{x}'Ax = \bar{\lambda}\bar{x}'x$.

Hence $(\lambda - \bar{\lambda})\bar{x}'x = 0$.

If $x = \begin{pmatrix} x_1 \\ x_2 \\ \cdot \\ \cdot \\ \cdot \\ \cdot \\ x_n \end{pmatrix} \neq 0$, then $\bar{x}'x = |x_1|^2 + \ldots\ldots + |x_n|^2 > 0$.

Thus $\lambda - \bar{\lambda} = 0$. Hence λ is real.

THEOREM 6.2.4 Let A be an $n \times n$ real symmetric matrix. Then the eigenvectors belonging to different eigenvalues are orthogonal.

PROOF

Suppose that x_i belongs to λ_i and x_j belongs to λ_j and $\lambda_i \neq \lambda_j$. Then

(1) $Ax_i = \lambda_i x_i$. Also $Ax_j = \lambda_j x_j$.

Transpose the latter:

(2) $x_j'A = \lambda_j x_j'$.

From (1) we get $x_j'Ax_i = \lambda_i x_j'x_i$ and from (2) we get $x_j'Ax_i = \lambda_j x_j'x_i$.

Thus $(\lambda_i - \lambda_j)x_j'x_i = 0$. But $\lambda_i - \lambda_j \neq 0$.

Hence $x_j'x_i = 0$.

This means that x_j and x_i are orthogonal.

THEOREM 6.2.5 (The Fundamental Theorem) Let A be an $n \times n$ real symmetric matrix. Then an $n \times n$ matrix P exists so that $P^{-1}AP$ is a **diagonal** matrix. We say that P transforms A and that $P^{-1}AP$ is the transform of A. Moreover P can be chosen to be an **orthogonal** matrix whose columns are the normalised eigenvectors of A.

The diagonal elements of $P^{-1}AP$ are the eigenvalues of A with their correct multiplicities as roots of $|A - \lambda I_n| = 0$.

PROOF

We give a proof by induction on n.

The result is trivial for $n=1$.

Suppose the result is true for any $(n-1) \times (n-1)$ matrix B which is real and symmetric.

Take A to be any $n \times n$ real symmetric matrix. Let λ and $\boldsymbol{x}$ be an eigenvalue and corresponding normalised eigenvector. Construct an orthogonal matrix $T = (\boldsymbol{x}_1, \boldsymbol{x}_2, \ldots\ldots, \boldsymbol{x}_n)$ as follows.

Take $\boldsymbol{x}_1 = \boldsymbol{x}$. For $\boldsymbol{x}_2$ take any normalised solution of $\boldsymbol{x}_1{}'\boldsymbol{x}_2 = 0$. For $\boldsymbol{x}_3$ take any normalised solution of the pair of equations: $\boldsymbol{x}_1{}'\boldsymbol{x}_3 = 0$ and $\boldsymbol{x}_2{}'\boldsymbol{x}_3 = 0$. For a general $\boldsymbol{x}_i$ take any normalised solution of the $i-1$ equations:

$$\boldsymbol{x}_1{}'\boldsymbol{x}_i = 0,\ \boldsymbol{x}_2{}'\boldsymbol{x}_i = 0, \ldots\ldots, \boldsymbol{x}_{i-1}'\boldsymbol{x}_i = 0,\ 2 \leqq i \leqq n.$$

In each case we have to solve **less than** n equations for the n elements

$$a_{1i}, a_{2i}, \ldots\ldots, a_{ni} \text{ of } \boldsymbol{x}_i = \begin{pmatrix} a_{1i} \\ \cdot \\ \cdot \\ \cdot \\ \cdot \\ a_{ni} \end{pmatrix}.$$

Thus there is **always** a solution $\boldsymbol{x}_i \neq \boldsymbol{0}$ by the remarks preceding theorem 4.16.3.

By construction, T is orthogonal.

Hence $T'T = I_n$.

Thus $T'\boldsymbol{x} = \begin{pmatrix} 1 \\ 0 \\ \cdot \\ \cdot \\ 0 \end{pmatrix}$, since $\boldsymbol{x}$ is the first column of T.

Since $A\boldsymbol{x} = \lambda\boldsymbol{x}$, we have $T'A\boldsymbol{x} = \lambda T'\boldsymbol{x} = \begin{pmatrix} \lambda \\ 0 \\ 0 \\ \cdot \\ \cdot \\ 0 \end{pmatrix}$.

Thus the first column of $T'AT$ *is* $\begin{pmatrix} \lambda \\ 0 \\ \cdot \\ \cdot \\ \cdot \\ 0 \end{pmatrix}$

Now A is symmetric, hence $A' = A$.

Hence $(T' A T)' = T' A' T'' = T' A T$.

Thus $T' A T$ is symmetric.

Hence $T' A T$ has the first row $(\lambda, 0, 0, \ldots\ldots, 0)$.

Thus $T' A T$ has the form:
$$\begin{pmatrix} \lambda & 0 \ldots\ldots 0 \ldots\ldots 0 \\ 0 & \\ \cdot & \\ \cdot & B \\ \cdot & \\ 0 & \end{pmatrix},$$

where B is a real symmetric $(n - 1) \times (n - 1)$ matrix.

By the induction hypothesis B can be transformed into diagonal form by means of an orthogonal matrix, S, say.

Consider $P' A P$, where $P = T \begin{pmatrix} 1 & 0 \ldots\ldots 0 \\ 0 & \\ \cdot & S \\ \cdot & \\ 0 & \end{pmatrix}$

We have $P' A P = \begin{pmatrix} 1 & 0 \ldots\ldots 0 \\ 0 & \\ \cdot & S' \\ \cdot & \\ 0 & \end{pmatrix} T' A T \begin{pmatrix} 1 & 0 \ldots\ldots 0 \\ 0 & \\ \cdot & S \\ \cdot & \\ 0 & \end{pmatrix} =$

$$= \begin{pmatrix} 1 & 0 \ldots\ldots 0 \\ 0 & \\ \cdot & S' \\ \cdot & \\ 0 & \end{pmatrix} \begin{pmatrix} \lambda & 0 \ldots\ldots 0 \\ 0 & \\ \cdot & B \\ \cdot & \\ 0 & \end{pmatrix} \begin{pmatrix} 1 & 0 \ldots\ldots 0 \\ 0 & \\ \cdot & S \\ \cdot & \\ 0 & \end{pmatrix} =$$

$$= \begin{pmatrix} \lambda & 0 \ldots\ldots 0 \\ 0 & \\ \cdot & S' B S \\ \cdot & \\ 0 & \end{pmatrix} = \text{a diagonal matrix.}$$

Now
$$\begin{pmatrix} 1 & 0 \cdots 0 \\ 0 & \\ \vdots & S \\ 0 & \end{pmatrix}$$
is orthogonal because

$$\begin{pmatrix} 1 & 0 \cdots 0 \\ 0 & \\ \vdots & S \\ 0 & \end{pmatrix}' \begin{pmatrix} 1 & 0 \cdots 0 \\ 0 & \\ \vdots & S \\ 0 & \end{pmatrix} =$$

$$= \begin{pmatrix} 1 & 0 \cdots 0 \\ 0 & \\ \vdots & S' \\ 0 & \end{pmatrix} \begin{pmatrix} 1 & 0 \cdots 0 \\ 0 & \\ \vdots & S \\ 0 & \end{pmatrix} = \begin{pmatrix} 1 & 0 \cdots 0 \\ 0 & \\ \vdots & S'S \\ 0 & \end{pmatrix}$$

$$= \begin{pmatrix} 1 & 0 \cdots 0 \\ 0 & \\ \vdots & I_{n-1} \\ 0 & \end{pmatrix} = I_n.$$

Moreover if U and W are orthogonal then so is $U W$. This follows from:

$$(U W)' (U W) = W' U' U W = W' I_n W = I_n.$$

Thus, as a product of two orthogonal matrices,

$$P = T \begin{pmatrix} 1 & 0 \cdots 0 \\ 0 & \\ \vdots & S \\ 0 & \end{pmatrix}$$
is itself orthogonal.

By induction on n, it follows that if A is an $n \times n$ real symmetric matrix then there exists an $n \times n$ orthogonal matrix P such that $P^{-1}A P$ is a diagonal matrix. (**Note** throughout that $M' = M^{-1}$ for M orthogonal).

It remains to show that the diagonal elements are the eigenvalues of A and the columns of P are the corresponding eigenvectors of A.

Now $P' A P = P^{-1} A P$ has the same eigenvalues as A by theorem 6.2.1.But the eigenvalues of

$$P^{-1} A P = \begin{pmatrix} \lambda_1 & 0 \ldots\ldots 0 \\ 0 & \lambda_2 \ldots\ldots \\ \cdot & \ddots \quad \cdot \\ \cdot & \quad \ddots \cdot \\ 0 \ldots\ldots\ldots & \lambda_n \end{pmatrix}$$

are precisely the diagonal elements. Thus $P^{-1} A P$ has the eigenvalues of A as its diagonal elements.

From

$$P^{-1} A P = \begin{pmatrix} \lambda_1 & 0 \ldots\ldots 0 \\ 0 & \lambda_2 \ldots\ldots \\ \cdot & \ddots \quad \cdot \\ \cdot & \quad \ddots \cdot \\ 0 \ldots\ldots\ldots & \lambda_n \end{pmatrix}$$

we have $$A P = P \begin{pmatrix} \lambda_1 & 0 \ldots\ldots 0 \\ 0 & \lambda_2 \ldots\ldots \\ \cdot & \ddots \quad \cdot \\ \cdot & \quad \ddots \cdot \\ 0 \ldots\ldots\ldots & \lambda_n \end{pmatrix}$$

Thus $$A (x_1, x_2, \ldots\ldots, x_n) = (x_1, x_2, \ldots\ldots, x_n) \begin{pmatrix} \lambda_1 & 0 \ldots\ldots 0 \\ 0 & \ddots \quad \cdot \\ \cdot & \quad \ddots \cdot \\ \cdot & \quad \ddots \cdot \\ 0 \ldots\ldots\ldots & \lambda_n \end{pmatrix}$$

Hence $$(Ax_1, Ax_2, \ldots\ldots, Ax_n) = (\lambda_1 x_1, \lambda_2 x_2, \ldots\ldots, \lambda_n x_n)$$

Thus $$Ax_i = \lambda_i x_i.$$

Hence the columns of P are eigenvectors of A, normalised and mutually orthogonal because P is known to be an orthogonal matrix.

Practical Method of Finding P so that $P'A P = P^{-1} A P$ is Diagonal

(1) **Case when eigenvalues of A are all distinct.**

By theorem 6.2.4, the eigenvectors are then mutually orthogonal. If we take the matrix $P = (x_1, x_2, \ldots\ldots, x_n)$, where the x_i are the normalised eigenvectors of A, then P is orthogonal and

$$A\,P = A\,(x_1, x_2, \ldots\ldots, x_n) = (\lambda_1 x_1, \lambda_2 x_2, \lambda_3 x_3, \ldots\ldots, \lambda_n x_n)$$

$$= (x_1, x_2, \ldots\ldots, x_n)\begin{pmatrix} \lambda_1 & 0 & \cdots & 0 \\ 0 & \lambda_2 & & \vdots \\ \vdots & & \ddots & \vdots \\ 0 & \cdots & \cdots & \lambda_n \end{pmatrix} = P\begin{pmatrix} \lambda_1 & 0 & \cdots & 0 \\ 0 & \ddots & & \vdots \\ \vdots & & \ddots & \vdots \\ 0 & \cdots & \cdots & \lambda_n \end{pmatrix}.$$

Thus $$P'A\,P = P^{-1}A\,P = \begin{pmatrix} \lambda_1 & 0 & \cdots & 0 \\ 0 & \lambda_2 & & \vdots \\ \vdots & & \ddots & \vdots \\ 0 & \cdots & \cdots & \lambda_n \end{pmatrix}$$

EXAMPLE 6.2.1

Let $$A = \begin{pmatrix} 10 & -14 & -10 \\ -14 & 7 & -4 \\ -10 & -4 & 19 \end{pmatrix}.$$

Characteristic equation of A is:

$$|A - \lambda I_3| = \lambda^3 - 36\,\lambda^2 + 81\,\lambda + 4374 = 0$$

Thus $(\lambda + 9)(\lambda - 18)(\lambda - 27) = 0.$

$\lambda_1 = -9.$ Normalised eigenvector is: $\frac{1}{3}\begin{pmatrix} 2 \\ 2 \\ 1 \end{pmatrix}$

$\lambda_2 = 18.$ " " " $\frac{1}{3}\begin{pmatrix} 1 \\ -2 \\ 2 \end{pmatrix}$

$\lambda_3 = 27.$ " " " $\frac{1}{3}\begin{pmatrix} -2 \\ 1 \\ 2 \end{pmatrix}.$

Take $$P = \tfrac{1}{3}\begin{pmatrix} 2 & 1 & -2 \\ 2 & -2 & 1 \\ 1 & 2 & 2 \end{pmatrix}.$$

$$P^{-1} = P' = \tfrac{1}{3}\begin{pmatrix} 2 & 2 & 1 \\ 1 & -2 & 2 \\ -2 & 1 & 2 \end{pmatrix} .$$

Hence $P^{-1}A\,P =$

$$\tfrac{1}{3}\begin{pmatrix} 2 & 2 & 1 \\ 1 & -2 & 2 \\ -2 & 1 & 2 \end{pmatrix}\begin{pmatrix} 10 & -14 & -10 \\ -14 & 7 & -4 \\ -10 & -4 & 19 \end{pmatrix}\tfrac{1}{3}\begin{pmatrix} 2 & 1 & -2 \\ 2 & -2 & 1 \\ 1 & 2 & 2 \end{pmatrix}$$

$$= \tfrac{1}{3}\begin{pmatrix} -18 & -18 & -9 \\ 18 & -36 & 36 \\ -54 & 27 & 54 \end{pmatrix}\tfrac{1}{3}\begin{pmatrix} 2 & 1 & -2 \\ 2 & -2 & 1 \\ 1 & 2 & 2 \end{pmatrix}$$

$$= \begin{pmatrix} -9 & 0 & 0 \\ 0 & 18 & 0 \\ 0 & 0 & 27 \end{pmatrix} .$$

Note that $|P| = -1$. If we want $|P| = 1$ we can multiply a column by -1 and take

$$P = \tfrac{1}{3}\begin{pmatrix} 2 & 1 & 2 \\ 2 & -2 & -1 \\ 1 & 2 & -2 \end{pmatrix}$$

Alternatively we can interchange suitable columns and take

$$P = \tfrac{1}{3}\begin{pmatrix} 1 & 2 & -2 \\ -2 & 2 & 1 \\ 2 & 1 & 2 \end{pmatrix} .$$

Then $P^{-1}A\,P = \begin{pmatrix} 18 & 0 & 0 \\ 0 & -9 & 0 \\ 0 & 0 & 27 \end{pmatrix}$,

with a change in the order of the eigenvalues down the diagonal.

(2) **Case when eigenvalues are repeated.**

Here, if the value λ is repeated r times in $|A - \lambda I_n| = 0$, we find r linearly independent eigenvectors belonging to the repeated root λ.

We take suitable linear combinations of these to obtain r normalised, mutually orthogonal eigenvectors all belonging to the root λ. These will be orthogonal to eigenvectors belonging to **other** roots of $|A - \lambda I_n| = 0$ by theorem 6.2.4. The existence of r linearly independent eigenvectors all belonging to the r-fold root λ, is guaranteed by theorem 6.2.5 in that the orthogonal transforming matrix P of that theorem has n linearly independent eigenvectors of A for its columns.

EXAMPLE 6.2.2

$$\text{Let } A = \begin{pmatrix} 5 & 2 & 2 \\ 2 & 2 & 1 \\ 2 & 1 & 2 \end{pmatrix}$$

Characteristic equation of A is:

$$|A - \lambda I_3| = \lambda^3 - 9\lambda^2 + 15\lambda - 7 = 0.$$

Thus $(\lambda - 1)^2(\lambda - 7) = 0.$

$$\lambda_1 = 7.$$

Normalised eigenvector is: $\dfrac{1}{\sqrt{6}}\begin{pmatrix} 2 \\ 1 \\ 1 \end{pmatrix}$.

$$\lambda_2 = \lambda_3 = 1.$$

Two eigenvectors of A belonging to the eigenvalue 1 are

$$\begin{pmatrix} -1 \\ 2 \\ 0 \end{pmatrix} \quad \text{and} \quad \begin{pmatrix} 0 \\ 1 \\ -1 \end{pmatrix}.$$

We note that the set of all vectors x satisfying $Ax = \lambda x$ is always a vector space. It is called the **eigenspace** belonging to λ. Thus our remarks preceding this example could be summed up by saying that if λ is an r-fold eigenvalue of a **symmetric** matrix A then the eigenspace belonging to λ has dimension r.

In our example the vectors

$$\begin{pmatrix} -1 \\ 2 \\ 0 \end{pmatrix} \quad \text{and} \quad \begin{pmatrix} 0 \\ 1 \\ -1 \end{pmatrix}$$

are linearly independent and form a basis for the eigenspace belonging to $\lambda=1$, which is 2-dimensional since $\lambda = 1$ is a double root. However these vectors are not orthogonal.

Take one of these vectors, say

$$\begin{pmatrix} 0 \\ 1 \\ -1 \end{pmatrix}, \text{ as one of the chosen eigenvectors for } \lambda = 1.$$

As the other, take $a\begin{pmatrix} -1 \\ 2 \\ 0 \end{pmatrix} + b\begin{pmatrix} 0 \\ 1 \\ -1 \end{pmatrix} = \begin{pmatrix} -a \\ 2a+b \\ -b \end{pmatrix}$.

Now this has to be orthogonal to $\begin{pmatrix} 0 \\ 1 \\ -1 \end{pmatrix}$. Thus $(2a + b) + b = 0$.

Hence $a + b = 0$.

Take $a = -1$ and $b = 1$.

Then $\begin{pmatrix} -a \\ 2a+b \\ -b \end{pmatrix} = \begin{pmatrix} 1 \\ -1 \\ -1 \end{pmatrix}$.

Now we normalise this vector and get $\frac{1}{\sqrt{3}}\begin{pmatrix} 1 \\ -1 \\ -1 \end{pmatrix}$.

Finally for $\lambda = 1$ we take as normalised eigenvectors: $\frac{1}{\sqrt{2}}\begin{pmatrix} 0 \\ 1 \\ -1 \end{pmatrix}, \frac{1}{\sqrt{3}}\begin{pmatrix} 1 \\ -1 \\ -1 \end{pmatrix}$.

Thus we have: $$P = \begin{pmatrix} \frac{2}{\sqrt{6}} & 0 & \frac{1}{\sqrt{3}} \\ \frac{1}{\sqrt{6}} & \frac{1}{\sqrt{2}} & \frac{-1}{\sqrt{3}} \\ \frac{1}{\sqrt{6}} & \frac{-1}{\sqrt{2}} & \frac{-1}{\sqrt{3}} \end{pmatrix}.$$

For this P we have $|P|=-1$. If it is required that $|P|=1$ then an interchange of columns will do the trick. Thus take:

$$P = \begin{pmatrix} \frac{2}{\sqrt{6}} & \frac{1}{\sqrt{3}} & 0 \\ \frac{1}{\sqrt{6}} & \frac{-1}{\sqrt{3}} & \frac{1}{\sqrt{2}} \\ \frac{1}{\sqrt{6}} & \frac{-1}{\sqrt{3}} & \frac{-1}{\sqrt{2}} \end{pmatrix} \quad \text{then } |P| = 1 \text{ and}$$

$$P^{-1}AP = P'AP =$$

$$\begin{pmatrix} \frac{2}{\sqrt{6}} & \frac{1}{\sqrt{6}} & \frac{1}{\sqrt{6}} \\ \frac{1}{\sqrt{3}} & \frac{-1}{\sqrt{3}} & \frac{-1}{\sqrt{3}} \\ 0 & \frac{1}{\sqrt{2}} & \frac{-1}{\sqrt{2}} \end{pmatrix} \begin{pmatrix} 5 & 2 & 2 \\ 2 & 2 & 1 \\ 2 & 1 & 2 \end{pmatrix} \begin{pmatrix} \frac{2}{\sqrt{6}} & \frac{1}{\sqrt{3}} & 0 \\ \frac{1}{\sqrt{6}} & \frac{-1}{\sqrt{3}} & \frac{1}{\sqrt{2}} \\ \frac{1}{\sqrt{6}} & \frac{-1}{\sqrt{3}} & \frac{-1}{\sqrt{2}} \end{pmatrix}$$

$$= \begin{pmatrix} 7 & 0 & 0 \\ 0 & 1 & 0 \\ 0 & 0 & 1 \end{pmatrix}.$$

6.3 CLASSIFICATION OF THE QUADRICS

Using our results of the previous section, we know that by a roation of axes corresponding to an orthogonal matrix P with $|P| = 1$, we can change the equation of the general quadric from the form:

$$a_{11}x_1^2 + a_{22}x_2^2 + a_{33}x_3^2 + 2a_{12}x_1x_2 + 2a_{13}x_1x_3 + 2a_{23}x_2x_3 +$$
$$+ 2b_1x_1 + 2b_2x_2 + 2b_3x_3 + c = 0,$$

to the form:

$$\lambda_1 x_1'^2 + \lambda_2 x_2'^2 + \lambda_3 x_3'^2 + 2u_1x_1' + 2u_2x_2' + 2u_3x_3' + d = 0,$$

where $\lambda_1, \lambda_2, \lambda_3$ are the eigenvalues of the symmetric matrix

$$A = \begin{pmatrix} a_{11} & a_{12} & a_{13} \\ a_{21} & a_{22} & a_{23} \\ a_{31} & a_{32} & a_{33} \end{pmatrix}, \text{ where } a_{ij} = a_{ji}.$$

The change of coordinates involved is:

$$\begin{pmatrix} x_1 \\ x_2 \\ x_3 \end{pmatrix} = P \begin{pmatrix} x_1' \\ x_2' \\ x_3' \end{pmatrix}, \text{ or } \quad x = Px'.$$

The columns of P are the eigenvectors of A, normalised and mutually orthogonal. We know that, since $P^{-1} = P'$,

$$\begin{pmatrix} x_1' \\ x_2' \\ x_3' \end{pmatrix} = P' \begin{pmatrix} x_1 \\ x_2 \\ x_3 \end{pmatrix}.$$

Moreover, if a_{rs} is the direction cosine of x_r' with respect to x_s then $P' = (a_{rs})$ as discussed in section 5.6.

Thus the rth row of P' consists of the direction cosines of x_r' with respect to the 'old' axes x_1, x_2, x_3.

Now the rth row of P' is the rth column of P.

Thus the direction cosines of the 'new' axes with respect to the 'old' are given by the eigenvectors of A, provided these are normalised and mutually orthogonal. (The direction cosines are in fact the components of the eigenvectors).

In view of the foregoing, we can now take the general quadric to be described by the equation:

$$\lambda_1 x_1^2 + \lambda_2 x_2^2 + \lambda_3 x_3^2 + 2u_1 x_1 + 2u_2 x_2 + 2u_3 x_3 + d = 0.$$

We now analyse the different cases that may arise.

Case 1. $\lambda_1, \lambda_2, \lambda_3$ **are all non-zero.**

We rearrange the equation to:

$$\lambda_1 \left(x_1 + \frac{u_1}{\lambda_1}\right)^2 + \lambda_2 \left(x_2 + \frac{u_2}{\lambda_2}\right)^2 + \lambda_3 \left(x_3 + \frac{u_3}{\lambda_3}\right)^2 = \frac{u_1^2}{\lambda_1} + \frac{u_2^2}{\lambda_2} + \frac{u_3^2}{\lambda_3} - d \equiv$$

$$\equiv d' \text{ (say)}.$$

Now transform the coordinates by **translation**, say:

$$x_1' = x_1 + \frac{u_1}{\lambda_1}$$

$$x_2' = x_2 + \frac{u_2}{\lambda_2}$$

$$x_3' = x_3 + \frac{u_3}{\lambda_3}$$

We get: $\lambda_1 x_1'^2 + \lambda_2 x_2'^2 + \lambda_3 x_3'^2 = d'$.

If $d' \neq 0$, we have:

$$x_1'^2 \Big/ \left(\frac{d'}{\lambda_1}\right) + x_2'^2 \Big/ \left(\frac{d'}{\lambda_2}\right) + x_3'^2 \Big/ \left(\frac{d'}{\lambda_3}\right) = 1.$$

This gives the following real possiblities, i.e. actual surfaces in 3-dimensional space.

(1) $\dfrac{x^2}{a^2} + \dfrac{y^2}{b^2} + \dfrac{z^2}{c^2} = 1$

(2) $\dfrac{x^2}{a^2} + \dfrac{y^2}{b^2} - \dfrac{z^2}{c^2} = 1$

(3) $\dfrac{x^2}{a^2} - \dfrac{y^2}{b^2} - \dfrac{z^2}{c^2} = 1$

where we are using x, y, z instead of x_1', x_2', x_3'.

If $d' = 0$, we have:

(4) $\lambda_1 x^2 + \lambda_2 y^2 + \lambda_3 z^2 = 0.$

This is a **quadric cone**. If (x, y, z) is on the surface, then so is (ax, ay, az) for any real number a. Thus straight lines through the origin lie on the surface.

The three quadrics (1), (2), and (3) are analogous to the ellipse and hyperbola in 2-dimensions.

They are called **central quadrics**. Their axes in the standard form given here are called **principal** axes. We note that the directions of the principal axes are given by the normalised, mutually orthogonal, eigenvectors of the matrix A.

The signs of $\lambda_1, \lambda_2, \lambda_3$, in conjunction with the sign of d' gives the type of quadric.

(1) is called an **ellipsoid** (2) is an **hyperboloid** of **one sheet** (3) is an **hyperboloid** of **two sheets.**

Figures 6.3.1 to 6.3.3 illustrate these surfaces.

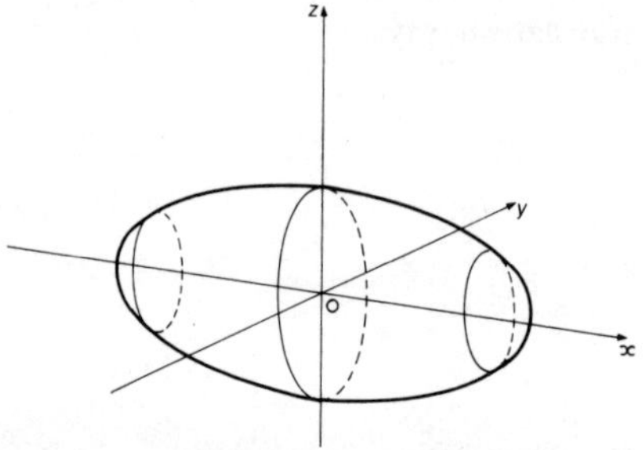

Figure 6.3.1 Ellipsoid.

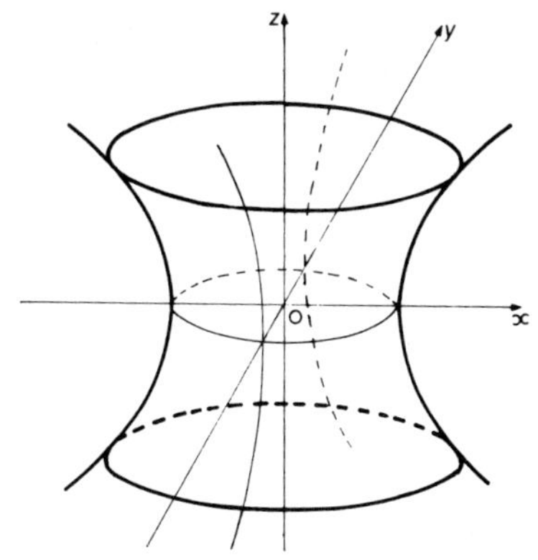

Figure 6.3.2 Hyperboloid of one sheet

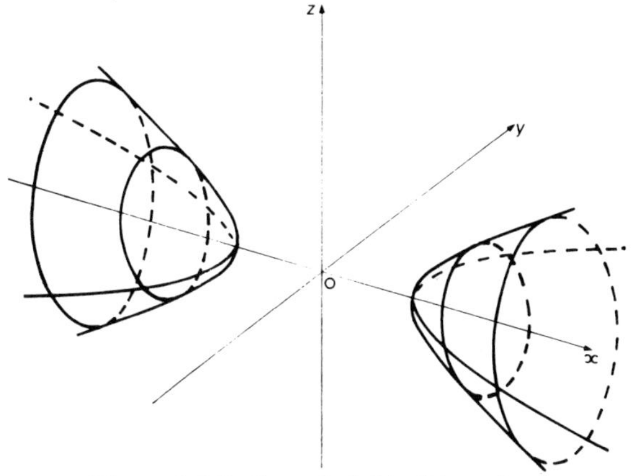

Figure 6.3.3 Hyperboloid of two sheets

Centre of Quadric.

When this type of quadric is in standard form, all chords through the origin are bisected there. This point of bisection is called the **centre** of the quadric. If we rotate axes and 'complete the square' as at the beginning of case 1 the centre is:

$$\left(\frac{-u_1}{\lambda_1}, \frac{-u_2}{\lambda_2}, \frac{-u_3}{\lambda_3}\right) .$$

Alternatively we can get the centre directly as follows.

Take the general equation of the (central) quadric:

$$a_{11} x_1^2 + a_{22} x_2^2 + a_{33} x_3^2 + 2a_{12} x_1 x_2 + 2a_{13} x_1 x_3 + 2a_{23} x_2 x_3 +$$

$$+ 2b_1 x_1 + 2b_2 x_2 + 2b_3 x_3 + c = 0.$$

If this is a central quadric referred to its centre as origin and (x_1, x_2, x_3) lies on the surface, then so does $(-x_1, -x_2, -x_3)$. Substituting both points into the general equation and subtracting one equation from the other we get: $b_1x_1 + b_2x_2 + b_3x_3 = 0$ for all (x_1, x_2, x_3) on the surface. This means that $b_1 = b_2 = b_3 = 0$. This shows that when the equation is referred to the centre as origin, the coefficients of the linear terms vanish.

Hence, to find the centre when we are given the general equation with linear terms,we change the coordinates by putting:

$$\begin{aligned} x_1 &= x_1' + a_1 \\ x_2 &= x_2' + a_2 \\ x_3 &= x_3' + a_3 \end{aligned}$$

and equate the coefficients of x_1', x_2', x_3' to zero.

The resulting equations give a_1, a_2, a_3 the coordinates of the centre. If the calculations are carried through for the general equation given above, the equations for the centre are:

$$\begin{aligned} a_{11}a_1 + a_{12}a_2 + a_{13}a_3 + b_1 &= 0 \\ a_{12}a_1 + a_{22}a_2 + a_{23}a_3 + b_2 &= 0 \\ a_{13}a_1 + a_{23}a_2 + a_{33}a_3 + b_3 &= 0 \end{aligned}$$

Note that in this case we have:

$$d' = \frac{u_1^2}{\lambda_1} + \frac{u_2^2}{\lambda_2} + \frac{u_3^2}{\lambda_3} - d = \lambda_1 a_1^2 + \lambda_2 a_2^2 + \lambda_3 a_3^2 - d,$$

using the fact that the centre is:

$$\left(\frac{-u_1}{\lambda_1}, \frac{-u_2}{\lambda_2}, \frac{-u_3}{\lambda_3}\right)$$

as given at the beginning of this discussion.

We return now to our analysis of the different types of quadric.

Case 2. One of $\lambda_1, \lambda_2, \lambda_3$ is 0, say $\lambda_3 = 0$.
Write the equation in the form:

$$\lambda_1\left(x_1 + \frac{u_1}{\lambda_1}\right)^2 + \lambda_2\left(x_2 + \frac{u_2}{\lambda_2}\right)^2 + 2u_3x_3 + d - \frac{u_1^2}{\lambda_1} - \frac{u_2^2}{\lambda_2} = 0.$$

Write $\quad 2u_3x_3 + d - \dfrac{u_1^2}{\lambda_1} - \dfrac{u_2^2}{\lambda_2} = 2u_3\left(x_3 + \left(d - \dfrac{u_1^2}{\lambda_1} - \dfrac{u_2^2}{\lambda_2}\right)\Big/2u_3\right),$

when $u_3 \neq 0$. Then by a suitable translation of axes the equation becomes:

(5) $\lambda_1 x^2 + \lambda_2 y^2 + 2u_3 z = 0,$

where again we use coordinates x, y, z rather than x_1', x_2', x_3'.

If $u_3 = 0$, we get:

(6) $\lambda_1 x^2 + \lambda_2 y^2 + d' = 0,$

when $d' = d - \dfrac{u_1^2}{\lambda_1} - \dfrac{u_2^2}{\lambda_2} \neq 0$, and

(7) $\lambda_1 x^2 + \lambda_2 y^2 = 0,$

when $d' = 0$.

Now (5) gives $\dfrac{x^2}{l} + \dfrac{y^2}{m} = 2z$ if λ_1, λ_2 have the same sign, and $\dfrac{x^2}{l} - \dfrac{y^2}{m} = 2z$ if λ_1, λ_2 have different signs. The former is called an **elliptic paraboloid** and the latter is called an **hyperbolic paraboloid**. They are illustrated in Figures 6.3.4 and 6.3.5.

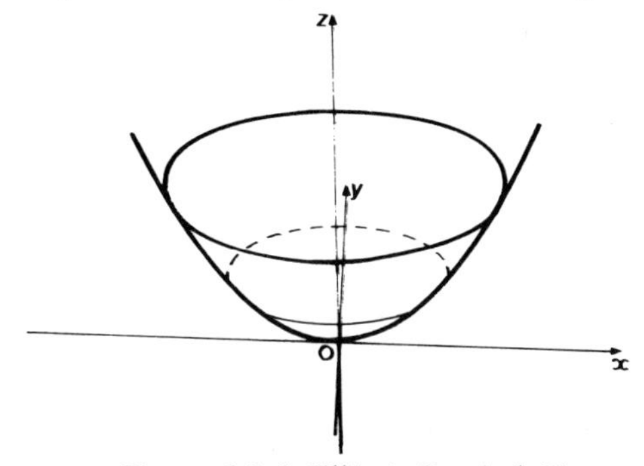

Figure 6.3.4 Elliptic Paraboloid

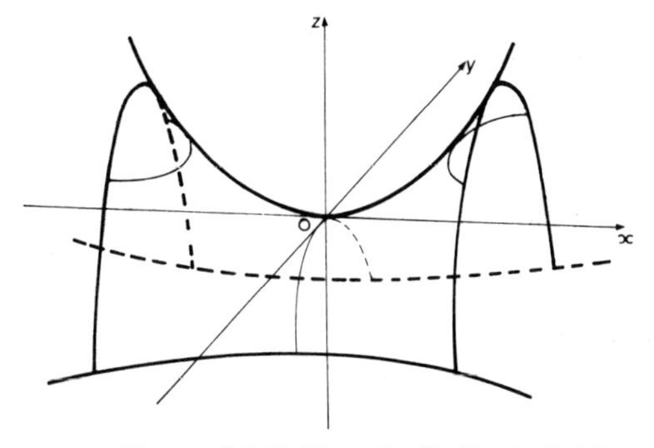

Figure 6.3.5 Hyperbolic Paraboloid

Equation (6) is a **hyperbolic cylinder** when λ_1 and λ_2 have different signs and an **elliptic cylinder** when λ_1 and λ_2 have the same sign.

These are shown in Figures 6.3.6 and 6.3.7.

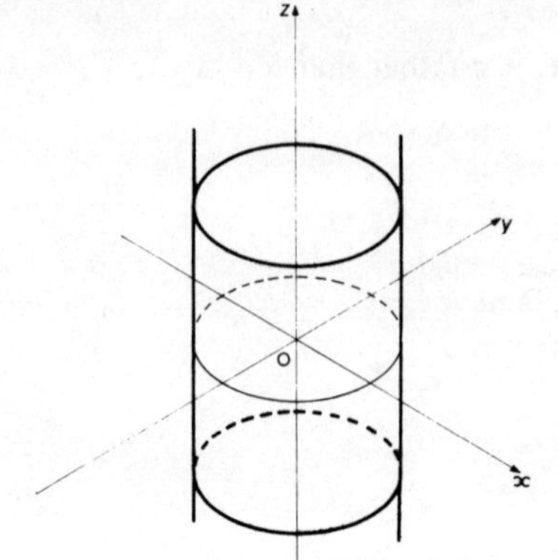

Figure 6.3.6 Elliptic Cylinder

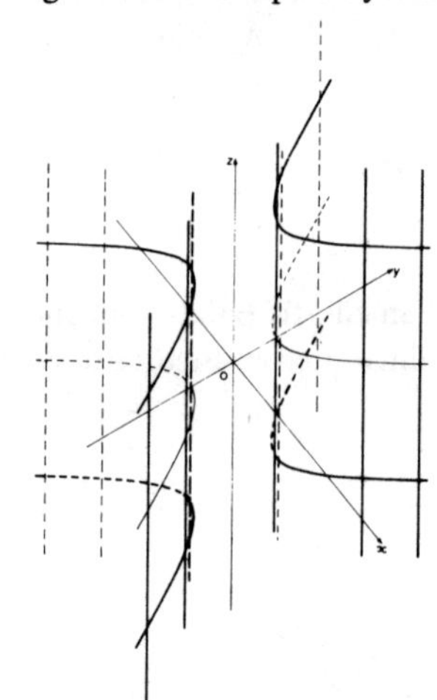

Figure 6.3.7 Hyperbolic Cylinder

Finally (7) gives two intersecting planes, imaginary if the signs of λ_1, λ_2 are the same and real if the signs of λ_1, λ_2 are different.

Case 3. Two of $\lambda_1, \lambda_2, \lambda_3$ are 0, say $\lambda_2 = \lambda_3 = 0$.

The equation becomes

$$\lambda_1 \left(x_1 + \frac{u_1}{\lambda_1}\right)^2 + 2u_2 x_2 + 2u_3 x_3 + d - \frac{u_1^2}{\lambda_1} = 0.$$

We change the coordinates to :

$$x = x_1 + \frac{u_1}{\lambda_1}$$

$$y = 2u_2 x_2 + 2u_3 x_3 + d - \frac{u_1^2}{\lambda_1}.$$

Note that y is in the plane of x_2 and x_3 so is perpendicular to x. To complete the change we take z to complete a right-hand system of axes with x and y.

The equation becomes:

(8) $x^2 = 2ky$, provided **not both** u_2 and u_3 vanish.

If $u_2 = u_3 = 0$, then the equation takes the form:

(9) $x^2 = k'$.

The former is a **parabolic cylinder** and the latter gives two parallel planes

$$x = \pm\sqrt{k'}$$

which are real when $k' \geqq 0$,

and coincident when $k' = 0$.

Figure 6.3.8 illustrates (8).

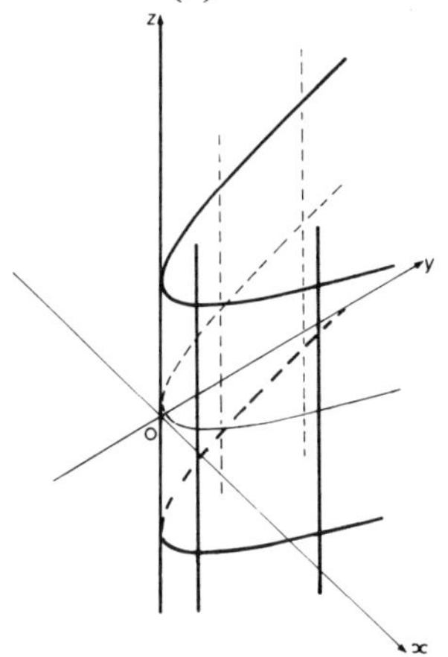

Figure 6.3.8 Parabolic Cylinder

6.4 RANK, SIGNATURE AND SYLVESTER'S LAW OF INERTIA

If we reduce a quadratic form by a **non-singular** transformation to a sum of squares in two different ways to get:

$$a_1 x_1^2 + a_2 x_2^2 + a_3 x_3^2 + \ldots\ldots + a_n x_n^2 \text{ and}$$

$$b_1 y_1^2 + b_2 y_2^2 + b_3 y_3^2 + \ldots\ldots + b_n y_n^2,$$

how are the a_i and b_j related?

Simple examples show that the a_i and b_j are not necessarily the same, even in a different order.

However something does remain unchanged as the following shows.

THEOREM 6.4.1 **(Sylvester's Law of Inertia)**

If $\sum_{i,j=1}^{n} a_{ij}x_ix_j$

is a real quadratic form in n variables and it is expressed as a sum of multiples of n real squares by a **non-singular** transformation **in any manner**, then the number of positive, the number of negative, and the number of zero coefficients are all independent of the mode of reduction.

PROOF

$$\text{Let } Q = \sum_{i,j=1}^{n} a_{ij}x_ix_j = \mathbf{x}'A\,\mathbf{x}, \text{ with } \mathbf{x} = \begin{pmatrix} x_1 \\ x_2 \\ . \\ . \\ x_n \end{pmatrix}.$$

First we deal with the number of zero coefficients.

Let $\mathbf{x} = T\mathbf{y}$, $|T| \neq 0$, so that:

$$\mathbf{x}'A\,\mathbf{x} = \mathbf{y}'(T'A\,T)\mathbf{y} = b_1y_1^2 + b_2y_2^2 + \ldots\ldots + b_ny_n^2.$$

$$\text{Then} \quad T'A\,T = \begin{pmatrix} b_1 & 0 \ldots\ldots 0 \\ 0 & b_2 \ldots\ldots 0 \\ . & . \quad\quad . \\ 0 & 0 \ldots\ldots b_n \end{pmatrix}.$$

Since T is non-singular, the rank of $T'A\,T =$ the rank of A, by theorem 4.16.1. Now the rank of $T'A\,T$ is the number of non-zero coefficients b_i. Hence the number of zero coefficients in any reduction to sum of squares form is (n – rank A), which is independent of the mode of reduction.

Now suppose that we have two reductions to sum of squares form. Let them be given by:

$$\mathbf{x} = T_1\,\mathbf{z} \text{ and } \mathbf{x} = T_2\,\mathbf{y}, \text{ so that}$$

$$x'Ax = \sum_{i=1}^{n} a_i z_i^2 \text{ and } x'Ax = \sum_{i=1}^{n} b_i y_i^2, \text{ where } a_1, a_2, \ldots, a_p$$

are positive and the remaining a_i are $\leqq 0$, and $b_1, b_2, \ldots\ldots, b_q$ are positive and the remaining b_j are $\leqq 0$.

Let $p < q$.

Put $z_1 = z_2 = \ldots\ldots = z_p = 0$ and $y_{q+1} = y_{q+2} = \ldots = y_n = 0$.

We have $\begin{pmatrix} z_1 \\ \cdot \\ \cdot \\ \cdot \\ z_n \end{pmatrix} = T_1^{-1} \begin{pmatrix} x_1 \\ \cdot \\ \cdot \\ \cdot \\ x_n \end{pmatrix}$ and $\begin{pmatrix} y_1 \\ \cdot \\ \cdot \\ \cdot \\ y_n \end{pmatrix} = T_2^{-1} \begin{pmatrix} x_1 \\ \cdot \\ \cdot \\ \cdot \\ x_n \end{pmatrix}$.

Thus we have:

$0 = z_1 =$ a **linear** combination of $x_1, x_2, \ldots\ldots, x_n = L_1(x_1, \ldots, x_n)$.

$0 = z_2 = L_2(x_1, \ldots, x_n)$

$0 = z_3 = L_3(x_1, \ldots, x_n)$

$0 = z_4 = L_4(x_1, \ldots, x_n)$

....

....

....

$0 = z_p = L_p(x_1, \ldots, x_n)$

$0 = y_{q+1} = L_{q+1}(x_1, \ldots, x_n)$

....

....

$0 = y_n = L_n(x_1, \ldots, x_n)$.

These give **less** than n linear homogeneous (0 on the left hand side) equations in the n variables $x_1, x_2, \ldots\ldots, x_n$. By the remarks preceding theorem 4.16.3, these equations have a non-zero solution $\mathbf{x}$.

For this solution $\mathbf{x}'A\,\mathbf{x} \leqq 0$ using the reduction

$$\mathbf{x}'A\,\mathbf{x} = \sum_{i=1}^{n} a_i z_i^2 . \quad \text{But also } \mathbf{x}'A\,\mathbf{x} = \sum_{i=1}^{n} b_i y_i^2 > 0,$$

because all the $b_1, b_2, \ldots\ldots, b_q$ are positive and not all the $y_1, y_2, \ldots\ldots, y_q$ can be zero or $\mathbf{x} = T\,\mathbf{y}$ would be zero. However we have obtained $\mathbf{x} \neq 0$.

This contradiction shows that $p \nless q$.

Similarly $q \nless p$. Thus $p = q$, and the number of positive coefficients is independent of the mode of reduction.

Now the number of non-zero coefficients in any reduction is the rank of A as shown at the beginning of this proof. This is independent of the mode of reduction. Thus the number of negative coefficients must also be independent of the mode of reduction.

This concludes the proof of the theorem.

Some definitions are now required.

DEFINITION 6.4.1

$$\text{Let } Q = \sum_{i,j=1}^{n} a_{ij} x_i x_j$$

be a quadratic form with associated symmetric $n \times n$ matrix $A = (a_{ij})$. Then the **rank** of Q is defined to be the rank of A, which, as in the above theorem, is the number of non-zero coefficients in any sum of squares form.

The **signature** of the form Q is defined to be (the number of positive coefficients) – (the number of negative coefficients) in any sum of squares form.

By the above theorem both the rank and the signature are invariants of the form Q.

DEFINITION 6.4.2 A quadratic form $Q = \mathbf{x}'A\,\mathbf{x}$ is said to be **positive definite** if and only if **any one** of the following equivalent conditions holds.

(1) Rank of $A = n =$ signature of Q.

(2) All the coefficients in the reduced form $\sum_{i=1}^{n} a_i z_i^2$ are positive.

(3) $\mathbf{x}'A\,\mathbf{x} > 0$ if $\mathbf{x} \neq \mathbf{0}$

$\mathbf{x}'A\,\mathbf{x}$ is said to be **negative definite** if and only if $-(\mathbf{x}'A\,\mathbf{x})$ is positive definite.

Often condition (2) serves as a good way of deciding if a given quadratic form is positive definite or not. However the following determinantal criterion sometimes proves useful particularly in theoretical investigations.

THEOREM 6.4.2 A necessary and sufficient condition that a quadratic form

$$Q = x'A\,x = \sum_{i,j=1}^{n} a_{ij}\,x_i\,x_j,\ a_{ij} = a_{ji},$$

should be positive definite is that the determinants:

$$a_{11},\quad \begin{vmatrix} a_{11} & a_{12} \\ a_{21} & a_{22} \end{vmatrix},\quad \begin{vmatrix} a_{11} & a_{12} & a_{13} \\ a_{21} & a_{22} & a_{23} \\ a_{31} & a_{32} & a_{33} \end{vmatrix},\ \ldots\ldots\ldots,\ |A|,$$

should all be positive.

PROOF

Put $Q_r = \sum_{i,j=1}^{r} a_{ij}\,x_i\,x_j.$

If Q_r could take a negative or zero-value for some non-zero

$$\boldsymbol{x} = \begin{pmatrix} x_1 \\ \cdot \\ \cdot \\ \cdot \\ x_r \end{pmatrix},$$

then taking the same $x_1, x_2, \ldots\ldots, x_r$ and putting $x_{r+1} = x_{r+2} = \ldots\ldots = x_n = 0$, we could make $Q = Q_n$ take a negative or zero value for a non-zero $\boldsymbol{x}$.

Thus if $Q = Q_n$ is positive definite then so is Q_r for $1 \leqq r \leqq n$.

Let $\boldsymbol{x} = T\boldsymbol{y}$ so that $Q_r = \sum_{i=1}^{r} b_i y_i^2$.

Then $T'A_r\,T = \begin{pmatrix} b_1 & 0\ldots\ldots 0 \\ 0 & b_2\ldots\ldots 0 \\ \cdot & \cdot \quad\quad \cdot \\ \cdot & \cdot \quad\quad \cdot \\ 0 & 0\ldots\ldots b_r \end{pmatrix}$, where $A_r = \begin{pmatrix} a_{11} & \ldots\ldots & a_{1r} \\ \cdot & & \cdot \\ \cdot & & \cdot \\ \cdot & & \cdot \\ a_{r1} & \ldots\ldots & a_{rr} \end{pmatrix}$

Thus $|T'A_r T| = b_1 b_2 \ldots . b_r > 0,$

because Q_r is positive definite.

Hence $|T|^2 |A_r| > 0.$

Thus $|A_r| > 0$ for $r = 1,2,3,4, \ldots . . \; n.$

Hence if $Q = Q_n$ is positive definite,

then $|A_r| > 0$ for $r = 1,2,3,4, \ldots \ldots , n.$

Conversely suppose $|A_r| > 0$ for $r = 1,2,3, \ldots \ldots , n.$

Go through the procedure for a completion of squares type reduction given in example 6.1.1. We have:

$$Q = Q_n = a_{11} x_1^2 + 2a_{12} x_1 x_2 + \ldots \ldots = a_{11} (x_1 + \frac{a_{12}}{a_{11}} x_2 + \ldots)^2$$

$$+ \ldots \ldots$$

Put $y_1 = x_1 + \dfrac{a_{12}}{a_{11}} x_2 + \ldots \ldots$

$y_i = x_i,$ for $i = 2,3,4, \ldots \ldots , n.$

Then $a_{11} = |A_1| > 0,$

and the transformation can be written:

$$y = \begin{pmatrix} 1 & \frac{a_{12}}{a_{11}} & \ldots & \ldots & \ldots & \frac{a_{1n}}{a_{11}} \\ 0 & 1 & 0 \ldots & \ldots & \ldots & 0 \\ 0 & 0 & 1 & 0 \ldots & \ldots & 0 \\ \cdot & \cdot & \cdot & \cdot & \cdot & \cdot \\ \cdot & \cdot & \cdot & \cdot & \cdot & \cdot \\ \cdot & \cdot & \cdot & \cdot & \cdot & \cdot \\ 0 & 0 & 0 \ldots & \ldots & \ldots & 1 \end{pmatrix} x = Tx, \text{ where } |T| = 1.$$

Thus if $P = T^{-1}$ then $|P| = 1$ and $x = Py.$

Hence $Q = Q_n = y' (P'AP) y = a_{11} y_1^2 + (\text{terms in } y_2, y_3, \ldots \ldots , y_n) =$

$= a_{11} y_1^2 + Q'$, say.

Thus $$P'AP = \begin{pmatrix} a_{11} & 0 \ldots\ldots 0 \\ 0 & \\ \cdot & B \\ \cdot & \\ 0 & \end{pmatrix}$$ and

$$|P'AP| = |P'||A||P| = |A|.$$

Hence $a_{11}|B| = |A|$.

Thus $|B| > 0$.

Now let P_r be the matrix of the first r rows and the first r columns of P.

Then $$\begin{pmatrix} a_{11} & 0 \\ 0 & B_{r-1} \end{pmatrix} = P_r' A_r P_r.$$

Thus $$\begin{vmatrix} a_{11} & 0 \\ 0 & B_{r-1} \end{vmatrix} = |P_r'||A_r||P_r|.$$

However $|P_r| = 1$.

Thus $$\begin{vmatrix} a_{11} & 0 \\ 0 & B_{r-1} \end{vmatrix} = |A_r|.$$

Hence $a_{11}|B_{r-1}| = |A_r| > 0$.

Thus $|B_{r-1}| > 0$ for $r - 1 = 1,2,3,4, \ldots\ldots, (n-1)$.

Thus the quadratic form Q' in $n - 1$ variables $y_2, \ldots\ldots, y_n$ with its associated matrix B satisfies the criterion of the theorem. An induction hypothesis assures us that this quadratic form is positive definite. Now $Q = Q_n = a_{11}\, y_1^2 + Q'$ and we are given $a_{11} > 0$. Thus Q must also be positive definite. Since the result is trivial for $n = 1$, induction on n completes the proof.

EXAMPLE 6.4.1

Let $Q_3 = 6x_1^2 + 5x_2^2 + 3x_3^2 - 2x_1x_2 + 6x_1x_3 - 6x_2x_3$.

Here $|A_1| = 6$, $|A_2| = \begin{vmatrix} 6 & -1 \\ -1 & 5 \end{vmatrix} = 29 > 0$, $|A| = 6 > 0$,

where $A = \begin{pmatrix} 6 & -1 & 3 \\ -1 & 5 & -3 \\ 3 & -3 & 3 \end{pmatrix}$. Reducing Q_3 according to the method

in the proof we have:

$$Q_3 = 6\left(x_1^2 - \tfrac{2}{6} x_1 x_2 + \tfrac{6}{6} x_1 x_3\right) + 5x_2^2 + 3x_3^2 - 6x_2 x_3$$

$$= 6\left(x_1 - \frac{x_2}{6} + \frac{3}{6} x_3\right)^2 + 5x_2^2 - \frac{x_2^2}{6} - \frac{6}{4} x_3^2 + x_2 x_3 + 3x_3^2$$

$$- 6x_2 x_3.$$

Put $\quad x_1' = x_1 - \dfrac{x_2}{6} + \dfrac{x_3}{2}$

$$x_2' = x_2$$

$$x_3' = x_3$$

then $\quad Q_3 - 6x_1'^2 = \dfrac{29}{6} x_2^2 + \dfrac{6}{4} x_3^2 - 5x_2 x_3.$

Put $\quad b_1 = 6,\ b_2 = 29/6,\ b_3 = 6/29.$

Then $\quad Q_3 - b_1 x_1'^2 = b_2 x_2^2 + \dfrac{6}{4} x_3^2 - 5x_2 x_3 = b_2\left(x_2 - \dfrac{30}{58} x_3\right)^2 +$

$+ \dfrac{6}{29} x_3^2.$

Thus $\quad Q_3 - b_1 x_1'^2 - b_2 y_2^2 = \dfrac{6}{29} x_3^2 = b_3 x_3^2.$

Hence $\quad Q_3 = b_1 x_1'^2 + b_2 y_2^2 + b_3 x_3^2 = |A_1| y_1^2 + \dfrac{|A_2|}{|A_1|} y_2^2 + \dfrac{|A|}{|A_2|} y_3^2,$

where $\quad y_2 = x_2 - \dfrac{30}{58} x_3, y_1 = x_1' = x_1 - \dfrac{x_2}{6} + \dfrac{x_3}{2}, y_3 = x_3' = x_3.$

Thus the transformation from:

$$Q_3 = 6x_1^2 + 5x_2^2 + 3x_3^2 - 2x_1x_2 + 6x_1x_3 - 6x_2x_3$$

to the sum of squares form: $\sum_{i=1}^{3} b_i y_i^2 = |A_1|\, y_1^2 + \frac{|A_2|}{|A_1|}\, y_2^2 + \frac{|A|}{|A_2|}\, y_3^2$

is: $\begin{pmatrix} y_1 \\ y_2 \\ y_3 \end{pmatrix} = T \begin{pmatrix} x_1 \\ x_2 \\ x_3 \end{pmatrix}$, where $T = \begin{pmatrix} 1 & \frac{-1}{6} & \frac{1}{2} \\ 0 & 1 & \frac{-30}{58} \\ 0 & 0 & 1 \end{pmatrix}$

6.5 SIMULTANEOUS REDUCTION OF TWO QUADRATIC FORMS

It is sometimes possible to change the coordinates or variables in such a way that two quadratic forms can be reduced simultaneously to sums of squares form. This reduction is not usually orthogonal.

The following theorem gives two occasions when the reduction is possible. One of these is particularly important for a physical application which we look at in a moment.

THEOREM 6.5.1 Two quadratic forms

$$Q_1 = \sum_{i,j=1}^{n} a_{ij} x_i x_j \quad \text{and} \quad Q_2 = \sum_{i,j=1}^{n} b_{ij} x_i x_j$$

can be transformed simultaneously to the forms:

$$Q_1 = \sum_{i=1}^{n} c_i y_i^2 \quad \text{and} \quad Q_2 = \sum_{i=1}^{n} d_i y_i^2$$

provided that **either** (1) the roots of the equation $|A - \lambda B| = 0$ are all distinct, where $A = (a_{ij})$ and $B = (b_{ij})$ are the symmetric matrices of Q_1 and Q_2, **or** (2) **one** of the two forms is positive definite.

PROOF

(1) Let the distinct roots of $|A - \lambda B| = 0$ be $\lambda_1, \lambda_2, \lambda_3, \ldots\ldots, \lambda_n$. Let

the corresponding solution vectors of the n linear homogeneous equations $(A - \lambda B)\,\boldsymbol{x} = \boldsymbol{0}$ be $\boldsymbol{x}_1, \boldsymbol{x}_2, \boldsymbol{x}_3, \ldots\ldots, \boldsymbol{x}_n$. These are sometimes called **right factors of zero** of the matrix $(A - \lambda B)$. Thus eigenvectors of A are the right factors of zero of the matrix $(A - \lambda I_n)$.

Then $\quad (A - \lambda_i B)\, x_i = \boldsymbol{0}.$

Thus $\quad A\, x_i = \lambda_i B\, x_i.$

Hence $\quad x_i' A = \lambda_i x_i' B$

since A and B are symmetric matrices. From these two equations, taking $\lambda_i \neq \lambda_j$, we get:

$$\lambda_i x_i' Bx_j = x_i' Ax_j = \lambda_j x_i' Bx_j$$

Hence $\quad x_i' Ax_j = x_i' Bx_j = 0.$

Put $\quad x_i' Ax_i = c_i,\ x_i' Bx_i = d_i$ and $T = (\boldsymbol{x}_1, \boldsymbol{x}_2, \ldots\ldots, \boldsymbol{x}_n)$.

Then $\quad T'A\,T = (x_s' Ax_t) = \begin{pmatrix} c_1 & 0 & \cdots & \cdots & 0 \\ 0 & c_2 & \cdots & \cdots & 0 \\ 0 & 0 & c_3 & \cdots & 0 \\ \vdots & \vdots & & & \vdots \\ 0 & 0 & \cdots & 0 & c_n \end{pmatrix}$ and

$$T'B\,T = \begin{pmatrix} d_1 & 0 & \cdots & 0 \\ 0 & d_2 & & \vdots \\ \vdots & & \ddots & \vdots \\ 0 & \cdots & \cdots & d_n \end{pmatrix}$$

Thus the change of variable given by $\begin{pmatrix} x_1 \\ \vdots \\ x_n \end{pmatrix} = T \begin{pmatrix} y_1 \\ \vdots \\ y_n \end{pmatrix}$

simultaneously reduces the two forms to:

$$\sum_{i=1}^{n} c_i y_i^2 \quad \text{and} \quad \sum_{i=1}^{n} d_i y_i^2 .$$

(2) We use induction on n. The result is trivial for $n = 1$. Assume the reduction is possible for forms in $(n - 1)$ variables. Take

$$Q_2 = \sum_{i,j=1}^{n} b_{ij} x_i x_j$$

to be the positive definite form.

Let λ be a root of $|A - \lambda B| = 0$.

Let $A\boldsymbol{x}_1 = \lambda B\boldsymbol{x}_1$ with $\boldsymbol{x}_1 \neq \boldsymbol{0}$.

Let $\boldsymbol{x}_2, \boldsymbol{x}_3, \ldots\ldots, \boldsymbol{x}_n$ be any $(n - 1)$ linearly independent solution vectors of the equation $\boldsymbol{x}_1{}'B\,\boldsymbol{x} = 0$. This is one equation in n variables. By theorem 4.16.2 the solution space has dimension $(n - 1)$. Thus the required $(n - 1)$ linearly independent solution vectors do exist.

Now $\boldsymbol{x}_1, \boldsymbol{x}_2, \ldots\ldots, \boldsymbol{x}_n$ will be linearly independent if and only if $\boldsymbol{x}_1$ does **not** lie in the solution space of $\boldsymbol{x}_1{}'B\,\boldsymbol{x} = 0$. This means that we want $\boldsymbol{x}_1{}'B\,\boldsymbol{x}_1 \neq 0$.

Now $\sum_{i,j=1}^{n} b_{ij} x_i x_j$ is positive definite.

Hence, since $\boldsymbol{x}_1 \neq \boldsymbol{0}$, we must have $\boldsymbol{x}_1{}'B\,\boldsymbol{x}_1 > 0$.

(Notice the use made here of $\sum_{i,j=1}^{n} b_{ij} x_i x_j$ being positive definite).

Put $T = (\boldsymbol{x}_1, \boldsymbol{x}_2, \ldots\ldots, \boldsymbol{x}_n)$. Then T is non-singular because its columns are linearly independent so its rank is n and so $|T| \neq 0$.

Now for $i > 1$ we have:

$$\boldsymbol{x}_1{}'A\,\boldsymbol{x}_i = \lambda \boldsymbol{x}_1{}'B\,\boldsymbol{x}_i = 0 \quad \text{and so also} \quad \boldsymbol{x}_i{}'A\,\boldsymbol{x}_1 = \boldsymbol{x}_i{}'B\,\boldsymbol{x}_1 = 0.$$

Thus $$T'AT = \begin{pmatrix} c & 0 \ldots\ldots 0 \\ 0 & \\ \cdot & \\ \cdot & C \\ \cdot & \\ \cdot & \\ 0 & \end{pmatrix}, \text{ where } c = \mathbf{x}_1'A\,\mathbf{x}_1 = \lambda \mathbf{x}_1' B\mathbf{x}_1$$

Similarly: $$T'BT = \begin{pmatrix} d & 0 \ldots\ldots 0 \\ 0 & \\ \cdot & \\ \cdot & D \\ \cdot & \\ \cdot & \\ 0 & \end{pmatrix}, \text{ where } d = \mathbf{x}_1' B\,\mathbf{x}_1 > 0.$$

As for the orthogonal reduction of one form, the induction hypothesis for the $(n-1) \times (n-1)$ matrices C and D enables us to deduce the required result for the forms in n variables. An appeal to induction then completes the proof.

Notes 6.5.1

(1) When repeated roots of $|A - \lambda B| = 0$ occur so that we can no longer use the first method above, yet one of the forms is positive definite so that the second part assures us that a simultaneous reduction is possible, we follow the procedure for the single reduction in similar circumstances. We take suitable linear combinations of the right factors of zero of $(A - \lambda B)$ which belong to the same repeated root. The linear combinations $\mathbf{x}_i$ must be chosen so that $\mathbf{x}_i' A\mathbf{x}_j = \mathbf{x}_i' B\mathbf{x}_j = 0$ when $i \neq j$, where $\mathbf{x}_i$ and $\mathbf{x}_j$ **both** belong to the **same** root λ.

(2) The case when $B = I_n$ in the above gives the single reduction, which, as we know, can be made orthogonal.

(3) When $\mathbf{x}'B\,\mathbf{x}$ is positive definite, we can choose our $\mathbf{x}_i$ for $T =$

$(\mathbf{x}_1, \mathbf{x}_2, \ldots\ldots, \mathbf{x}_n)$ so that

$$\sum_{i,j=1}^{n} b_{ij} x_i x_j \text{ reduces to } y_1^2 + y_2^2 + \ldots\ldots + y_n^2.$$

For example just choose $\mathbf{x}_1$ so that $\mathbf{x}_1' B\mathbf{x}_1 = 1$ etc.

(4) When one of the forms is positive definite we can reduce it to

$$\sum_{i=1}^{n} y_i^2, \text{ and the other form to, } \sum_{i,j=1}^{n} c_{ij}\, y_i\, y_j, \text{ say.}$$

Then an orthogonal reduction of

$$\sum_{i,j=1}^{n} c_{ij}\, y_i\, y_j \text{ leaves } \sum_{i=1}^{n} y_i^2 \text{ still a sum of squares.}$$

For example if $\boldsymbol{y} = T\boldsymbol{z}$, with T orthogonal, reduces

$$\sum_{i,j=1}^{n} c_{ij}\, y_i\, y_j \text{ to } \sum_{i=1}^{n} d_i z_i^2, \text{ then it reduces } \sum_{i=1}^{n} y_i^2 \text{ to}$$

$$\boldsymbol{z}'T'T\boldsymbol{z} = \boldsymbol{z}'\boldsymbol{z} = \sum_{i=1}^{n} z_i^2, \text{ which is still a sum of squares.}$$

This gives a two stage reduction of the two forms to sums of squares. By combining the two transformations we get a single reduction.

EXAMPLE 6.5.1 We give one example which is an important physical application. A simple pendulum with 'bob' of mass m hangs from a point of support 0 by a string of negligible mass and length l.

An exactly similar pendulum hangs from the bob of the first pendulum. The axis $0x$ is horizontal and the axis $0y$ is vertically downward. Generalised coordinates are the minimum number of parameters that determine the given physical configuration. In our case, we take the two angles θ and ϕ. (For further information on generalised coordinates and related matters including Lagrange Equations to be used shortly, the reader is referred to any book on analytical dynamics, e.g. '*Principles of Mechanics*' by Synge and Griffith (McGraw-Hill)).

We denote the potential energy of the system by V and the kinetic energy by T. We arrange that $V = 0$ at the equilibrium position of the system.

Figure 6.5.1 illustrates the situation.

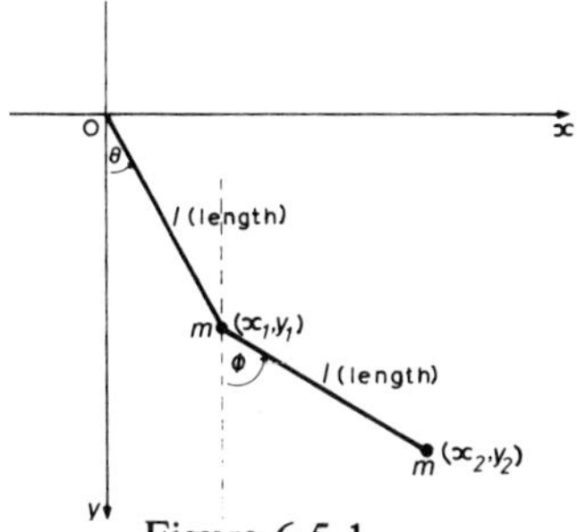

Figure 6.5.1

The calculations go as follows.

$$V = 3mgl - mgl\cos\theta - mgl(\cos\theta + \cos\phi).$$

For small displacements $\cos\theta = 1 - \theta^2/2$, etc.

Thus $V = (mgl/2)\,G$, where $G = 2\,\theta^2 + \phi^2$.

$$T = \frac{m}{2}(\dot{x}_1^2 + \dot{y}_1^2) + \frac{m}{2}(\dot{x}_2^2 + \dot{y}_2^2),\ \dot{x}_1 \text{ denotes } \frac{dx_1}{dt} \text{ etc.}$$

$$\begin{aligned}
x_1 &= l\sin\theta & \dot{x}_1 &= l(\cos\theta)\,\dot{\theta}\\
y_1 &= l\cos\theta & \dot{y}_1 &= -l(\sin\theta)\,\dot{\theta}\\
x_2 &= l\sin\theta + l\sin\phi & \dot{x}_2 &= l(\cos\theta)\,\dot{\theta} + l(\cos\phi)\,\dot{\phi}\\
y_2 &= l\cos\theta + l\cos\phi & \dot{y}_2 &= -l(\sin\theta)\,\dot{\theta} - l(\sin\phi)\,\dot{\phi}
\end{aligned}$$

Thus for small displacements when $\cos\theta = 1 - \theta^2/2$ and $\sin\theta = \theta$, we have:

$$\dot{x}_1^2 + \dot{y}_1^2 = l^2\,\dot{\theta}^2,\ \dot{x}_2^2 + \dot{y}_2^2 = l^2\,\dot{\theta}^2 + l^2\,\dot{\phi}^2 + 2l^2\,\dot{\theta}\,\dot{\phi}.$$

$$\begin{aligned}
\text{Hence}\quad T &= \frac{m}{2}(2l^2\,\dot{\theta}^2 + l^2\,\dot{\phi}^2 + 2l^2\,\dot{\theta}\,\dot{\phi})\\
&= (ml^2/2)\,F, \text{ where } F = 2\dot{\theta}^2 + 2\dot{\theta}\,\dot{\phi} + \dot{\phi}^2
\end{aligned}$$

is positive definite, because kinetic energy is > 0 unless $\dot{\theta} = \dot{\phi} = 0$.

Put $A = \begin{pmatrix} 2 & 1 \\ 1 & 1 \end{pmatrix}$ and $B = \begin{pmatrix} 2 & 0 \\ 0 & 1 \end{pmatrix}$,

then A is the matrix associated with F and B is associated with G.

We consider $(A - \lambda B)$. We have:

$$|A - \lambda B| = \begin{vmatrix} 2-2\lambda & 1 \\ 1 & 1-\lambda \end{vmatrix} = 0.$$

Thus $\lambda_1 = 1 + \frac{\sqrt{2}}{2}$, $\lambda_2 = 1 - \frac{\sqrt{2}}{2}$.

$$(A - \lambda_1 B) = \begin{pmatrix} -\sqrt{2} & 1 \\ 1 & -\frac{\sqrt{2}}{2} \end{pmatrix} \quad \text{and } x_1 = \begin{pmatrix} 1 \\ \sqrt{2} \end{pmatrix}.$$

$$(A - \lambda_2 B) = \begin{pmatrix} \sqrt{2} & 1 \\ 1 & \frac{\sqrt{2}}{2} \end{pmatrix} \quad \text{and } x_2 = \begin{pmatrix} 1 \\ -\sqrt{2} \end{pmatrix}.$$

Take $P = \begin{pmatrix} 1 & 1 \\ \sqrt{2} & -\sqrt{2} \end{pmatrix}$. Then put $\begin{pmatrix} \theta \\ \phi \end{pmatrix} = P \begin{pmatrix} \alpha \\ \beta \end{pmatrix}$

Thus $\theta = \alpha + \beta$, $\quad \dot{\theta} = \dot{\alpha} + \dot{\beta}$

$\phi = \sqrt{2}(\alpha - \beta)$, $\quad \dot{\phi} = \sqrt{2}(\dot{\alpha} - \dot{\beta})$

and $\alpha = \frac{1}{2}\left(\theta + \frac{\phi}{\sqrt{2}}\right)$, $\quad \beta = \frac{1}{2}\left(\theta - \frac{\phi}{\sqrt{2}}\right)$.

These latter coordinates α and β relative to which T and V become sums of squares are called NORMAL COORDINATES.

F becomes $4\lambda_1 \dot{\alpha}^2 + 4\lambda_2 \dot{\beta}^2$ and

G becomes $4\alpha^2 + 4\beta^2$ when expressed in normal coordinates.

Under certain circumstances that hold in this problem the equations of motion of a dynamical system described by generalised coordinates can be expressed in a very neat form called **Lagrange equations.** In our case they assume the form:

$$\frac{d}{dt}\left(\frac{\partial T}{\partial \dot{\alpha}}\right) - \frac{\partial T}{\partial \alpha} = \frac{-\partial V}{\partial \alpha} \text{ and } \frac{d}{dt}\left(\frac{\partial T}{\partial \dot{\beta}}\right) - \frac{\partial T}{\partial \beta} = \frac{-\partial V}{\partial \beta},$$

where $T = ml^2/_2\,(4\lambda_1 \dot{\alpha}^2 + 4\lambda_2 \dot{\beta}^2)$

and $V = mgl/_2\,(4\alpha^2 + 4\beta^2)$.

On reduction these give:

$$\ddot{\alpha} + \left(\frac{g}{l\lambda_1}\right) \alpha = 0 \quad \text{and} \quad \ddot{\beta} + \left(\frac{g}{l\lambda_2}\right) \beta = 0.$$

These are simple harmonic motion type equations with general solutions:

$$\begin{cases} \alpha = \alpha_o \cos\left(\sqrt{\dfrac{g}{l\lambda_1}}\ t + a\right) \\ \beta = \beta_o \cos\left(\sqrt{\dfrac{g}{l\lambda_2}}\ t + b\right) \end{cases} .$$

The so-called **normal modes** of vibration are given by taking:

(1) $\beta_o = 0, \quad \alpha_o \neq 0$

(2) $\beta_o \neq 0, \quad \alpha_o = 0.$

The corresponding periods of the normal modes are:

$$\frac{2\pi}{\sqrt{g/l\lambda_1}} \quad \text{and} \quad \frac{2\pi}{\sqrt{g/l\lambda_2}} .$$

Notice that the general solution is a linear combination of the normal modes. (Compare Fourier series for periodic functions).

Also note that if T and V are **not** sums of squares the equations do not take the simple form above but both equations involve both the generalised coordinates with a corresponding increase in the difficulty of solving. Even more benefit is seen by working in normal coordinates when there is an additional forcing vibration imposed on the system.

In practice, there are other, sometimes more convenient methods of finding normal coordinates. This is so in particular when certain special conditions hold.

We shall not pursue the topic here, but content ourselves with the observation that the method we have used is probably as good as any to **prove** that the simultaneous reduction to sums of squares form **can** be effected.

It should also be pointed out that, in practice, even in the case of the single reduction, it is not easy to find eigenvalues and the corresponding eigenvectors. We have always chosen convenient numbers so that the solution has been fairly easy to obtain by exact methods. This problem is of great importance in many practical applications.

For a further discussion of this question, the interested reader is referred to the introductory text: '*Computational Methods for Matrix Eigenproblems*' by Gourlay and Watson (Wiley 1973).

EXERCISES

1. Reduce the following quadratic forms to 'sum of squares' form by a non-singular transformation as in example 6.1.1 by completing the square.

(a) $x^2 + 2y^2 + 3z^2 + 6xy + 8yz$

(b) $x^2 + 5y^2 + 3z^2 + 2xy + 4xz$

(c) $xy + zy + zx$

In each case express the transformation in the form:

$$\begin{pmatrix} x' \\ y' \\ z' \end{pmatrix} = T \begin{pmatrix} x \\ y \\ z \end{pmatrix},$$

and find whether or not T is orthogonal.

2. Reduce the following quadratic forms to 'sum of squares' form by an **orthogonal** transformation.

(a) $x^2 + 5y^2 + 3z^2 + 8yz - 8xz$

(b) $5x^2 + 6y^2 + 7z^2 - 4xy + 4yz$

(c) $5x^2 + 11y^2 - 2z^2 + 12xz + 12yz$

3. Find an **orthogonal** transformation T with $|T| = 1$, that is a rotation of axes, so that the quadric

$$3x_1^2 + 3x_2^2 + 5x_3^2 + 2x_1x_2 + 2x_1x_3 + 2x_2x_3 = 1$$

becomes $$\lambda_1 y_1^2 + \lambda_2 y_2^2 + \lambda_3 y_3^2 = 1$$

in the new axes, where $\boldsymbol{y} = T\boldsymbol{x}$ with

$$\boldsymbol{y} = \begin{pmatrix} y_1 \\ y_2 \\ y_3 \end{pmatrix}, \text{ and } \quad \boldsymbol{x} = \begin{pmatrix} x_1 \\ x_2 \\ x_3 \end{pmatrix}.$$

What type of quadric is this?

4. Show that

$$2x_1^2 + 2x_2^2 - x_3^2 - 4x_2 x_3 - 4x_1 x_3 + 8x_1 x_2 = 1$$

is the equation of a quadric of revolution (i.e. can be obtained by rotating a plane curve through 360 degrees about an appropriate axis), and find the direction of its axis of symmetry. (Here, as in all questions, the axes $0x_1 x_2 x_3$ are the usual mutually orthogonal straight line Cartesian axes).

5. Find a rotation of axes so that the quadric:

$$7x_1^2 + 5x_2^2 + 3x_3^2 - 8x_1 x_2 + 8x_2 x_3 = 1$$

takes the form $\lambda_1 y_1^2 + \lambda_2 y_2^2 + \lambda_3 y_3^2 = 1.$

What type of quadric is this?

6. Find the coordinates of the **centre** of the central quadric:

$$4x^2 + 9y^2 + 6z^2 + 4yz + 8x + 40y + 20z + 34 = 0.$$

By finding the eigenvalues and eigenvectors of a certain 3 x 3 matrix, show that the quadric is an ellipsoid and find the directions and the lengths of its principal axes.

7. Reduce **simultaneously** to sums of squares form:

$$x^2 + 5y^2 + 3z^2 - 6xy - 2xz + 6yz \quad \text{and}$$

$$2x^2 + 6y^2 + 5z^2 + 10yz + 2xz.$$

8. Find which of the following forms are positive definite. Also find the rank and signature of each form.

(a) $3x^2 + 5y^2 + 7z^2 - 6xy + 8yz - 4xz$

(b) $x^2 + y^2 + z^2 + 2xy + 2yz + 2xz$

(c) $x^2 + y^2 + z^2 - 2xy - 2yz - 2xz$

(d) $2x^2 + 2y^2 + z^2 + 2xz + 2yz.$

9. In a dynamical problem on the small oscillations of a system with two degrees of freedom using generalised coordinates q_1 and q_2, the kinetic energy is given by:

$$T = (m/2)(3\dot{q}_1^2 + 3\dot{q}_2^2 - 2\dot{q}_1\dot{q}_2)$$

and the potential energy is given by:

$$V = mg\,(5q_1^2 + 5q_2^2 + 2q_1 q_2).$$

By a simultaneous reduction of the two quadratic forms T and V, find a set of normal coordinates.

By use of the Lagrange equations for the system, find the periods of the normal modes.

In the above

$$\dot{q}_i = \frac{dq_i}{dt},$$

t is time, m is mass, g is the acceleration due to gravity, all measured in appropriate units. The Lagrange equations for the system using generalised coordinates x_1 and x_2 are:

$$\frac{d}{dt}\left(\frac{\partial T}{\partial \dot{x}_i}\right) - \frac{\partial T}{\partial x_i} = \frac{-\partial V}{\partial x_i}$$

for $i = 1$ and 2.

CHAPTER 7

Factorisation and Euclidean Domains

7.1 INTRODUCTION

In this chapter we return to arithmetic as the source of our ideas. We are all familiar with the following three facts about factorisation in the ring of integers **Z**.

(1) If n is an integer not equal to 0 or ± 1, then $n = p_1 p_2 p_3 \ldots\ldots p_r$, where the p_i are primes **not necessarily distinct**. If also $n = q_1 q_2 q_3 \ldots\ldots q_s$, with the q_i primes, then $r = s$, and with a suitable change of order, $p_i = u_i q_i$, where $u_i = \pm 1$.

Note that 1 and -1 are the only elements of **Z** which have multiplicative inverses in **Z**. Such elements are called **units** of **Z**. Let S be the set of **all** the units in an **arbitrary** ring with an identity 1.

Let $a, b \in S$.

Then $(ab)^{-1} = b^{-1} a^{-1}$.

Thus $ab \in S$.

Also $(a^{-1})^{-1} = a$.

Since the associative rule is inherited from the ring, it now follows that S is a **group** under the multiplication in the ring. We call S the **group** of **units** of the ring. A group generated by one element is called a **cyclic** group. If it is finite it looks like $\{e, a, a^2, \ldots\ldots, a^{n-1}\}$, where $a^n = e$, the identity of the group. Here a has **order** n. If it is infinite it looks like $\{e, a^{\pm 1}, a^{\pm 2}, \ldots\ldots\}$ and is isomorphic to **Z** under +. For the definition of group and order of an element, see definition 4.3.1 and example 4.3.1 together with the remarks preceding the latter.

For **Z** the group of units is cyclic generated by -1, i.e. $\{-1, (-1)^2\} = \{-1, 1\}$ and is of order 2.

(2) Euclid's algorithm holds in **Z**. If $a, b \in \mathbf{Z}$, with $b \neq 0$, then $a = qb + r$, where $0 < |r| < |b|$ or $r = 0$.

(3) Every pair $a, b \in \mathbf{Z}$ has a greatest common divisor (G.C.D) $d > 0$. d is characterised by the property that d divides a, written $d \mid a$, $d \mid b$ and if $c \mid a$ and $c \mid b$, then $c \mid d$. Moreover we have $d = A\,a + B\,b$, for some $A, B \in \mathbf{Z}$.

These ideas turn out to be important in proving results in other branches of mathematics. For example, let G be a group. Let $a \in G$. Let r be a positive integer, and let the order of a be n. (See example 4.3.1 and remarks preceding it).

Then a^r also has order n if and only if (r, n), the greatest common divisor of r and n, is 1.

We sometimes describe the latter by saying that r and n are **coprime**.

To prove the above result we argue as follows.

Let $\quad (r, n) = 1.$

Then by (3) above, $\; 1 = R\,r + N\,n \;$ for some $\; R, N \in \mathbf{Z}.$

Now $\quad (a^r)^n = (a^n)^r = e^r = e.$

Thus the order $\; a^r \leqq n$. But $\; a = a^1 = a^{Rr+Nn} =$

$$= (a^r)^R\,(a^n)^N = (a^r)^R\,e^N = (a^r)^R\,e = (a^r)^R.$$

If the order of a^r is m, then

$$a^m = (a^r)^{Rm} = ((a^r)^m)^R = e^R = e.$$

Thus the order $a \leqq m$. We now have:

$$\text{order } a^r \leqq \text{order } a \quad \text{and} \quad \text{order } a \leqq \text{order } a^r.$$

Thus we have: $\quad$ order a = order a^r.

Conversely if $(r, n) = d > 1$, then let $n = sd$ and $r = td$. We have:

$$(a^r)^s = a^{tds} = a^{tn} = (a^n)^t = e^t = e.$$

Thus the order $a^r \leqq s < n$.

Hence, if the order of $a^r = n$, then $(r, n) = 1$.

Polynomial Rings.

Perhaps not so well known is the fact that the ring $F[x]$ of all polynomials in x with coefficients from the field F has similar properties to those listed above for **Z**.

In the case of $F[x]$, **prime** is just an irreducible polynomial i.e. a polynomial that cannot be factorised into a product of factors lieing in $F[x]$.

For example $x^2 + 1 \in \mathbf{Q}[x]$ may be factorised as the product $(x - i)(x + i)$ in the ring $\mathbf{C}[x]$ but cannot be factorised in $\mathbf{Q}[x]$.

Thus $x^2 + 1$ is **irreducible** or **prime** in $\mathbf{Q}[x]$ but is **reducible** in $\mathbf{C}[x]$.

The units in $F[x]$ are just the non-zero elements of the field F. These are the non-zero polynomials of degree 0; the constant polynomials.

The group of units is the multiplicative group of non-zero elements of F, that is $F - \{0\}$, usually denoted by F^*.

The role of modulus $|\ |$ of property (2) is played by the degree of the polynomial.

As usual to avoid proving the same result afresh in each new context, we formulate the foregoing ideas in an abstract form and prove the required results once and for all. We then apply these general results to each special case as it arises. The type of ring, which captures just those properties discussed above, is called a Euclidean domain. The vital property turns out to be (2) with a suitable generalisation of modulus of an integer or degree of a polynomial. The other properties may be deduced, as we shall see.

7.2 EUCLIDEAN DOMAINS

Formally we have:

DEFINITION 7.2.1 A Euclidean domain D is an integral domain (a commutative ring with an identity and with no divisors of zero i.e. $ab = 0$ implies that either $a = 0$ or $b = 0$) together with a function $\delta : D - \{0\} \to \{0,1,2,3, \ldots, n, \ldots\}$ satisfying :

(1) for $a, b \in D, a \neq 0, b \neq 0$, we have $\delta(ab) \geqq \delta(a)$.

(2) (Euclid's algorithm) for any $a, b \in D, b \neq 0$, there exists an expression $a = qb + r$, where either $r = 0$ or $\delta(r) < \delta(b)$.

We shall prove that all the results given previously for **Z** also hold for D. In fact, since **Z** is a special case of D, our proofs will also be proofs of the factorisation results for integers.

However, before giving the proofs we need several more definitions.

DEFINITION 7.2.2 A non-unit p is said to be a prime in D if whenever $p = ab$ for some $a, b \in D$, then either a or b is a unit of D.

DEFINITION 7.2.3 We say that a divides b written $a \mid b$ if and only if $b = ac$ for some $c \in D$.

DEFINITION 7.2.4 Let $a \neq 0, b \neq 0$ be elements of D. We say that a and b are **associates** if and only if $a = ub$ for some unit $u \in D$. This is equivalent to $a \mid b$ and $b \mid a$.

First we prove the equivalence of the statements in the above definition.

PROOF

Suppose that $a \mid b$ and $b \mid a$. Then $b = ac$ and $a = bd$ for suitable $c, d \in D$. Thus $a = bd = acd$. Hence $a(1 - cd) = 0$.

Now D is an integral domain. Thus $a = 0$ or $1 - cd = 0$. But we are assuming that $a \neq 0$. Hence $1 - cd = 0$. Thus $cd = 1$. Hence c and d are units. Thus $a = db$ with d a unit. Hence a and b are associates.

Conversely, let $a = ub$, with u a unit. Then immediately $b \mid a$. Now u^{-1} exists, hence $u^{-1} a = b$. Thus $a \mid b$.

DEFINITION 7.2.5 An ideal I of D is a subset of D with the properties:

(1) if $a, b \in I$, then $a + b \in I$

(2) if $d \in D$, and $a \in I$, then $da \in I$.

For example, when $D = \mathbf{Z}$, $\{3n : n \in \mathbf{Z}\} = \langle 3 \rangle$ is an ideal generated by 3.

DEFINITION 7.2.6 A principal ideal I of D is an ideal I consisting of all multiples of a single element of D. We say that I is **generated** by that element. Thus $I = \{ra : r \in D\}$, for some **fixed** $a \in D$. We write $I = \langle a \rangle$.

In the example above $\{3n : n \in \mathbf{Z}\} = \langle 3 \rangle$ is a principal ideal of $\mathbf{Z}$.

DEFINITION 7.2.7 A prime ideal I is an ideal with the property that if $ab \in I$, then either $a \in I$ or $b \in I$.

EXAMPLE 7.2.1 We give a few examples of prime ideals. For any **commutative** ring R the whole ring R is an example of a prime ideal.

The **zero ideal** $\{0\}$ is a prime ideal in any integral domain. In $\mathbf{Z}$, any ideal generated by a prime (i.e. all multiples of a prime) is a prime ideal. For example $\{\ldots -21, -14, -7, 0, 7, 14, 21, \ldots\} = \langle 7 \rangle$. If $ab \in \langle 7 \rangle$, then $ab = r7$ for some $r \in \mathbf{Z}$. Thus $7 \mid ab$. Hence $7 \mid a$ or $7 \mid b$. Thus $a = 7s$ or $b = 7t$. But then $a \in \langle 7 \rangle$ or $b \in \langle 7 \rangle$.

DEFINITION 7.2.8 A **principal ideal domain** is an integral domain in which **every** ideal is a principal ideal.

The following theorem tells us that Euclidean domains are principal ideal domains. The converse is false.

THEOREM 7.2.1 In any Euclidean domain D every ideal I is a principal ideal.

PROOF

Let I be any ideal of D. If $I = <0>$, then I is principal. Suppose $I \neq <0>$.

Choose $a \in I$, $a \neq 0$, such that $\delta(a)$ is smallest possible. Let b be any other element in I.

By property (2) of D, we have $b = qa + r$ with $r = 0$ or $\delta(r) < \delta(a)$.

If $r \neq 0$, then $\delta(r) < \delta(a)$ and $r = b - qa \in I$ because I is an ideal. ($a \in I$, hence $(-q)\,a \in I$, and also $b \in I$, thus $b + (-q)a \in I$).

However this contradicts the choice of a with the smallest $\delta(a)$. Thus $r = 0$ and $b = qa$.

Thus $I \subset <a>$. Now $<a> \subset I$ because I is an ideal. We use property (2). Altogether $I = <a>$. The theorem follows.

We can now prove property (3) for $\mathbf{Z}$.

THEOREM 7.2.2 If a and b are any two elements of D, then there exists a G.C.D $d = (a, b)$. Moreover d can be written in the form $d = A\,a + B\,b$ for some A and B from D.

PROOF

Let I be the ideal generated by a and b. This means that

$$I = \{ra + sb : r, s \in D\} = <a, b>.$$

By the previous theorem I is principal. Hence $I = <d>$, say. But then $a = ud$ and $b = vd$, for some $u, v \in D$. Thus $d \mid a$ and $d \mid b$.

Moreover $d \in I$, hence $d = A\,a + B\,b$ for some A and B in D. Now if $d' \mid a$ and $d' \mid b$, then $d' \mid (A\,a + B\,b)$. Thus $d' \mid d$. Hence d is a G.C.D of a and b, and $d = A\,a + B\,b$.

Note. If d_1 and d_2 are two G.C.D's of a and b, then $d_1 \mid d_2$ and $d_2 \mid d_1$. Thus d_1 and d_2 are associates.

Conversely, if d is a G.C.D of a and b, then ud is also a G.C.D of a and b, where u is a unit of D.

We pause for a moment in our development of the theory in order to give a method for finding $d = (a,b)$ and expressing it as $d = A\,a + B\,b$.

Method for d and Expressing as $d = A\,a + B\,b$.

Let $\delta(a) \leqq \delta(b)$. By property (2) of D we have:

$$b = q_1 a + a_1,\ \delta(a_1) < \delta(a)$$

$$a = q_2 a_1 + a_2,\ \delta(a_2) < \delta(a_1)$$

$$a_1 = q_3a_2 + a_3, \delta(a_3) < \delta(a_2)$$

$$\cdots\cdots\cdots\cdots\cdots\cdots\cdots$$

$$\cdots\cdots\cdots\cdots\cdots\cdots\cdots$$

$$a_{s-2} = q_sa_{s-1} + a_s, \delta(a_s) < \delta(a_{s-1}),$$

and so on.

Because the $\delta(a_i)$ for all i are positive integers or 0, this process must stop after a finite number of steps with:

$$a_{t-1} = q_{t+1}\, a_t.$$

Now all numbers $b, a, a_1, \ldots\ldots, a_t$ are of the form $u\, a + v\, b$. Moreover a_t divides $a_{t-1}, a_{t-2}, \ldots\ldots a_{s-1}, \ldots\ldots, a, b$. Thus a_t is the G.C.D of a and b.

Later in this section we shall work some specific examples of this technique. For the moment we return to our development of the theory.

THEOREM 7.2.3 If in a Euclidean domain D, b is **not** an associate of a, but $a \mid b$, then $\delta(a) < \delta(b)$.

PROOF

We have: $a = qb + r$, $r \neq 0$, $\delta(r) < \delta(b)$. Also $b = ac$ for some $c \in D$. Thus $r = a - qb = a - qac = a(1 - qc)$. Hence $a \neq 0$, $(1 - qc) \neq 0$ and therefore $\delta(a) \leqq \delta(r)$ by property (1) of D.

However $\delta(r) < \delta(b)$. Thus $\delta(a) < \delta(b)$.

THEOREM 7.2.4 In a Euclidean domain D, every element $b \neq 0$, which is **not** a unit, is a product of prime elements: $b = p_1\, p_2\, p_3 \ldots\ldots p_r$, where the primes p_i are not necessarily distinct.

PROOF

By induction on $\delta(b)$. Let $\delta(b) = 0$.

If $b = ac$, where neither a nor c is a unit, then b is not an associate of a, so $\delta(a) < \delta(b)$.

However $\delta(b) = 0$, hence this is impossible. Thus b is a prime.

Now suppose that the theorem is true for all elements x with $\delta(x) < \delta(b)$. If b is a prime, there is nothing to prove. If $b = ac$, with a and c not associates of b, then $\delta(a) < \delta(b)$ and $\delta(c) < \delta(b)$.

By the induction hypothesis, a and c are products of primes. Hence $b = ac$ is a product of primes.

The theorem follows by induction on $\delta(b)$.

THEOREM 7.2.5 In a Euclidean domain D, if a prime element p divides ab, then $p \mid a$ or $p \mid b$.

PROOF

Suppose $p \nmid a$. Then, since p is prime, the G.C.D of p and a is a unit u. Thus $u = x\,a + y\,p$ for some $x, y \in D$. Hence $u^{-1}\,u = u^{-1}\,x\,a + u^{-1}\,y\,p$.

Thus $1 = A\,a + P\,p$ for some $A, P \in D$. But then $b = bA\,a + b\,P\,p = A(ab) + {} + P\,(bp)$.

Now $p \mid ab$ and $p \mid bp$. Thus $p \mid b$.

THEOREM 7.2.6 If $p \in D$ is a prime element and $p \mid (a_1\,a_2\,a_3\,.\,.\,.\,.\;a_r)$ then $p \mid a_i$ for some i, where $1 \leqq i \leqq r$.

PROOF

The result is true for $r = 2$ by the theorem above and for $r = 1$ trivially. If $p \mid a_r$ then the required result follows.

Suppose that $p \nmid a_r$, then $p \mid (a_1\,a_2\,.\,.\,.\,.\,.\,a_{r-1})$ by the above theorem. Appeal to an induction hypothesis shows that $p \mid a_i$ for some $1 \leqq i \leqq (r-1)$.

Thus always $p \mid a_i$ for some $1 \leqq i \leqq r$.

The required result follows by induction on r.

THEOREM 7.2.7 Let $b = p_1\,p_2\,p_3\,.\,.\,.\,.\,.\,p_r = q_1\,q_2\,q_3\,.\,.\,.\,.\,.\,q_s$ be two factorisations of b as a product of primes in the Euclidean domain D, where $b \neq 0$ and b is **not** a unit.

Then $r = s$, and with suitable rearrangement $p_i = u_i\,q_i$, where u_i is a unit.

PROOF

For $r = 1$, $b = p_1$ is a prime and $r = s = 1, p_1 = q_1$.

We prove the result in general by induction on r.

We have $p_1 \mid q_1\,q_2\,q_3\,.\,.\,.\,.\,.\,q_s$. Thus $p_1 \mid q_i$ for some i by the above theorem. By rearrangement we can make $p_1 \mid q_1$. (Here we are renumbering the q_i).

Thus $q_1 = u_1\,p_1$ with u_1 a unit, because q_1 is a prime.

Hence $p_1\,p_2\,p_3\,.\,.\,.\,.\,.\,p_r = u_1\,p_1\,q_2\,.\,.\,.\,.\,.\,q_s$.

Thus $p_1(p_2\,p_3\,.\,.\,.\,p_r - u_1\,q_2\,q_3\,.\,.\,.\,.\,.\,q_s) = 0$.

Now because D is an integral domain and $p_1 \neq 0$, we must have:

$$p_2\;p_3\;.\,.\,.\,.\,.\;p_r - u_1\,q_2\,q_3\,.\,.\,.\,.\,.\,q_s = 0.$$

Thus $p_2\,p_3\,.\,.\,.\,.\,.\,p_r = u_1\,q_2\,q_3\,.\,.\,.\,.\,.\,q_s$. Put $u_1\,q_2 = q_2{}'$.

Then $q_2{}'$ is a prime. The induction hypothesis now operates to show that $r - 1 = s - 1$. Hence $r = s$. Moreover the p's and q's are the same apart from order and multiplication by units.

The theorem follows.

EXAMPLE 7.2.2 **Z** with $\delta(n) = |n|$ and **Q** $[x]$ with $\delta(f(x)) =$ the degree of $f(x)$ are Euclidean domains. The primes in **Q** $[x]$ are the irreducible polynomials like $x^2 + 1$.

We have: $24 = 3 \times 2 \times 2 \times 2 = (-3) \times (-2) \times 2 \times 2$

$= 2 \times (-3) \times 2 \times (-2)$ etc,

and $x^4 - 1 = (x-1)(x+1)(x^2+1) =$

$= (1/2)(x-1)(x+1)\ 2(x^2+1) =$

$= (-1/3)(x-1)(x^2+1)(-3x-3)$ etc.

Now let us find the G.C.D of 782 and 256.

$$782 = 3(256) + 14$$

$$256 = 18(14) + 4$$

$$14 = 3(14) + 2$$

$$4 = 2(2).$$

Thus 2 is the G.C.D of 782 and 256. Moreover

$$2 = 14 - 3(4)$$

$$= 782 - 3(256) - 3(256 - 18(14))$$

$$= 782 - 3(256) - 3(256) + 54(782 - 3(256)).$$

Thus $2 = 55(782) - 168(256)$.

Finally we find the G.C.D of $x^4 - 1$ and $x^3 + 2x^2 + 2x + 1$ in **Q** $[x]$.

$$x^4 - 1 = (x-2)(x^3 + 2x^2 + 2x + 1) + (2x^2 + 3x + 1)$$

$$x^3 + 2x^2 + 2x + 1 = \tfrac{1}{4}(2x+1)(2x^2 + 3x + 1) + \tfrac{3}{4}(x+1)$$

$$2x^2 + 3x + 1 = (2x+1)(x+1).$$

Thus the G.C.D is: $x + 1$ or, in fact, $k(x + 1)$ for any $k \in \mathbf{Q}$. Moreover:

$$3(x+1) = 4(x^3+2x^2+2x+1) - (2x+1)(2x^2+3x+1) =$$

$$= 4(x^3+2x^2+2x+1) - (2x+1)\{(x^4-1) - (x-2)(x^3+2x^2+2x+1)\}$$

$$= (2x^2-3x+2)(x^3+2x^2+2x+1) - (2x+1)(x^4-1)$$

$$= F(x) \cdot (x^3+2x^2+2x+1) + G(x)(x^4-1).$$

Note that in **Z** it is possible to find the G.C.D by using the factorisation of the numbers concerned into products of primes. However in other Euclidean domains such as $F[x]$ it is a very difficult problem to decide when an element actually is a prime. Thus the technique just illustrated is particularly useful.

7.3 THE RING OF GAUSSIAN INTEGERS

$$\mathbf{Z}[i] = \{x + iy : x, y \in \mathbf{Z}\}$$

is called the ring of Gaussian integers. It is a sub ring of **C**, hence is an integral domain.

We would like to prove that $\mathbf{Z}[i]$ is a Euclidean domain, then all the properties of such rings would hold for $\mathbf{Z}[i]$.

First we define $\delta : \mathbf{Z}[i] - \{0\} \to \{0,1,2,\ldots,n,\ldots\}$

by $\qquad \delta(a) = |a|^2 = x^2 + y^2$, where $a = x + iy \neq 0$.

Note that when $a \neq 0$, $\delta(a) \geqq 1$. Thus, for a, b in $\mathbf{Z}[i] - \{0\}$, we have:

$$\delta(ab) = \delta(a)\,\delta(b) \geqq \delta(a).$$

We want to show that there exist q and r in $\mathbf{Z}[i]$ so that $a = qb + r$, with either $r = 0$ or $\delta(r) < \delta(b)$.

Find a/b as a complex number. Let it be represented by P on the Argand Diagram shown in Figure 7.3.1.

Take the nearest lattice point Q to P. Then Q defines a Gaussian integer q. If $P \neq Q$, then $a = qb + r$, where r is defined to be $r = a - qb$, $r \neq 0$.

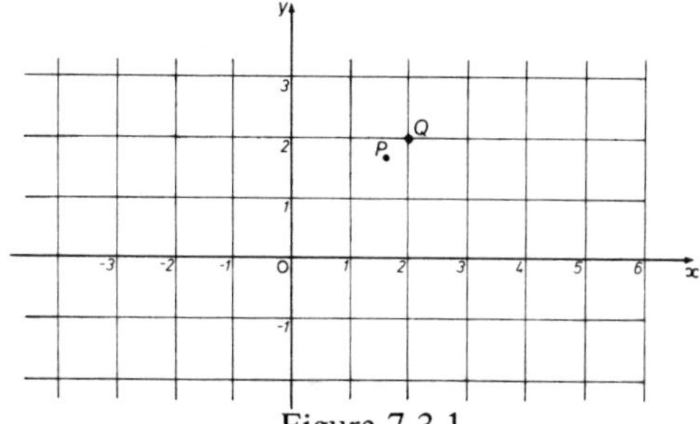

Figure 7.3.1

From the diagram, $|a/b - q| \leqq \sqrt{1/2}$. Thus $|a/b - q|^2 \leqq 1/2$. But $(a - qb)/b = r/b$. Thus $|r/b|^2 \leqq 1/2$. Hence $|r|^2 \leqq (1/2)|b|^2$. That is: $\delta(r) \leqq (1/2)\ \delta(b)$. Hence: $\delta(r) < \delta(b)$.

If $P = Q$, then $a = qb$ and $r = 0$. Thus $a = qb + r$ with either $r = 0$ or $\delta(r) < \delta(b)$.

It follows that $\mathbf{Z}[i]$ is a Euclidean domain.

Units.

The units of $\mathbf{Z}[i]$ are $\pm 1, \pm i$. The proof of this is given as an exercise for the reader. (See exercise 7.4).

EXAMPLE 7.3.1 Find the G.C.D d of the Gaussian integers $5 + i$ and $3 - i$ and express d in the form $d = A(5 + i) + B(3 - i)$, where A and B are Gaussian integers.

Solution.

Take $a = 5 + i$ and $b = 3 - i$. We have: $(5 + i)/(3 - i) = 7/5 + (4/5)\,i$. The nearest Gaussian integer is: $1 + i$. This can be seen from Figure 7.3.2.

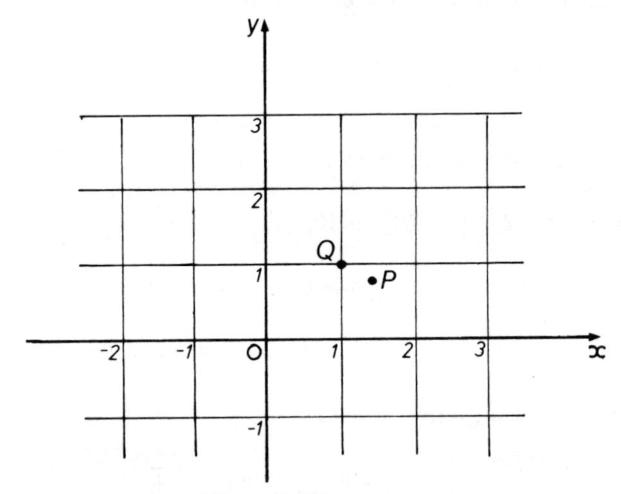

Figure 7.3.2

Thus $(5 + i) = (1 + i)(3 - i) + (1 - i)$.

We have $\delta(1 - i) = |1 - i|^2 = 2 < 10 = \delta(3 - i)$.

Now $b/r = (3 - i)/(1 - i) = 2 + i$, which is already Gaussian. Thus:

$(3 - i) = (2 + i)(1 - i)$. The remainder this time is 0 because

$q = (2 + i)$ is Gaussian.

Thus $r = (1 - i)$ is the required G.C.D of $5 + i$ and $3 - i$. We have:

$$(1 - i) = 1(5 + i) + (-(1 + i))(3 - i).$$

Hence $(1 - i) = A(5 + i) + B(3 - i)$ with

$A = 1$ and

$B = -(1 + i)$, both in $\mathbf{Z}[i]$ as required.

The fact that $\mathbf{Z}[i]$ is Euclidean can be used to prove results in the theory of numbers.

For example, if p is a prime with $p = 4n + 1$, then $p = a^2 + b^2$, for some integers a and b.

7.4 APPLICATION TO NUMBER THEORY

As an application of the ideas discussed in this chapter let us consider the proof of the theorem due to Fermat mentioned in the preceding paragraph.

We do this through a sequence of theorems starting with a result about abelian groups which is of some interest for its own sake.

THEOREM 7.4.1 Let A be a finite multiplicative **abelian** group with elements $a_1, a_2, a_3, \ldots\ldots, a_n$. Let $b_1, b_2, b_3, \ldots\ldots, b_r$ be those elements of A of order 2; that is $b_i^2 = e$, $b_i \neq e$, where e is the identity of A, $i = 1,2,3,4, \ldots\ldots, r$. Then $a_1 a_2 a_3 a_4 \ldots\ldots a_n = b_1 b_2 b_3 \ldots\ldots b_r$.

PROOF

Let $c_1^{\pm 1}, c_2^{\pm 1}, c_3^{\pm 1}, \ldots\ldots c_t^{\pm 1}$ be those elements of A whose orders are **greater** than 2. Then

$$A = \{e, b_1, b_2, \ldots\ldots, b_r, c_1, c_2, \ldots\ldots, c_t, c_1^{-1}, c_2^{-1}, \ldots\ldots, c_t^{-1}\}.$$

$$\begin{aligned} \text{Thus } a_1 a_2 a_3 a_4 \ldots\ldots a_n &= e\, b_1 b_2 b_3 \ldots\ldots b_r\, c_1 c_1^{-1} c_2 c_2^{-1} \ldots\ldots c_t c_t^{-1} \\ &= b_1 b_2 b_3 b_4 \ldots\ldots b_r\, e\, e \ldots\ldots e \\ &= b_1 b_2 b_3 b_4 \ldots\ldots b_r, \text{ as required.} \end{aligned}$$

The set $\mathbf{Z}_m = \{\bar{0}, \bar{1}, \bar{2}, \bar{3}, \ldots\ldots, \overline{(m - 1)}\}$ of residue classes modulo a positive integer m was introduced in Example 2.14.1, shown to form a ring under a suitable definition of addition and multiplication in section 3.1, and further discussed in Exercises 3.1 and 3.2.

We now prove

THEOREM 7.4.2 Let p be a prime integer. Then $\mathbf{Z}_p = \{\overline{0}, \overline{1}, \overline{2}, \overline{3}, \ldots\ldots, \overline{(p-1)}\}$ is a **field**.

PROOF

From section 3.1 we know that $\mathbf{Z}_p$ is a commutative ring. We have to show that if $\overline{a} \neq \overline{0}$, $\overline{a} \in \mathbf{Z}_p$, then $\overline{a}$ has a multiplicative inverse, i.e. there exists $\overline{A}$ so that $\overline{A}\,\overline{a} = \overline{1}$.

Let $\overline{a} \neq \overline{0}$. Then $p \nmid a$. Hence $(p,a) = 1$.

Now $\mathbf{Z}$ is a **Euclidean domain**, hence $1 = P\,p + A\,a$ for certain integers P and A. Thus

$$\begin{aligned}\overline{1} &= \overline{P\,p + A\,a} = \overline{P}\,\overline{p} + \overline{A}\,\overline{a}\\ &= \overline{P}\,\overline{0} + \overline{A}\,\overline{a}\\ &= \overline{A}\,\overline{a}.\end{aligned}$$

Thus $\overline{a}$ does have a multiplicative inverse $\overline{A}$. Hence $\mathbf{Z}_p$ is a field.

We can now deduce a famous result in the theory of numbers. It is usual to write $a \equiv b$, and say that a is **congruent** to b modulo p, if and only if $\overline{a} = \overline{b}$, that is if and only if $p \mid (a - b)$.

THEOREM 7.4.3 (Wilson's Theorem) If p is a prime integer, then $(p-1)! \equiv$ $\equiv -1$ (modulo p), or in our other notation, $\overline{(p-1)!} = \overline{(-1)}$.

PROOF

By theorem 7.4.2 $\mathbf{Z}_p^* = \mathbf{Z}_p - \{\overline{0}\} = \{\overline{1}, \overline{2}, \overline{3}, \ldots\ldots, \overline{(p-1)}\}$ is a finite multiplicative abelian group.

Now $(\overline{a})^2 = \overline{1}$ if and only if $p \mid (a^2 - 1)$, where $1 \leqq a \leqq (p-1)$.

But $a^2 - 1 = (a+1)(a-1)$. Hence $p \mid (a^2 - 1)$ if and only if

$p \mid (a+1)$ or $p \mid (a-1)$, by theorem 7.2.5.

The only possibilities are $a = p - 1$ or $a = 1$. Thus $\overline{a} = \overline{1}$ or $\overline{a} = \overline{(p-1)}$. Hence the only element of order 2 in $\mathbf{Z}_p^*$ is $\overline{(p-1)}$.

By theorem 7.4.1 we have $\overline{1} \cdot \overline{2} \cdot \overline{3} \ldots\ldots \overline{(p-1)} = \overline{(p-1)}$.

Now $\overline{(p-1)} = \overline{(-1)}$, because $p \mid ((p-1) - (-1))$.

Hence $\overline{1} \cdot \overline{2} \cdot \overline{3} \ldots\ldots \overline{(p-1)} = \overline{(-1)}$.

Thus $\overline{(p-1)!} = \overline{(-1)}$.

This is $(p-1)! \equiv -1$ (modulo p), as required.

Using Wilson's theorem, we have

THEOREM 7.4.4 If $p = 4n + 1$ is a prime integer, then there exists an integer x so that

$$x^2 \equiv -1 \text{ (modulo } p).$$

PROOF

Put $x = (2n)! = 1 \cdot 2 \cdot 3 \ldots\ldots (2n)$.

Now x has an even number of factors, hence we may write

$$x = (-1)(-2)(-3) \ldots\ldots (-2n).$$

Thus $x^2 = (1\ 2\ 3 \ldots (2n))((-1)(-2) \ldots\ldots (-2n))$.

Now $p - k \equiv -k$ modulo p, hence

$$p - 1 \equiv -1, p - 2 \equiv -2, p - 3 \equiv -3, \text{ and so on.}$$

Hence $x^2 \equiv 1 \cdot 2 \cdot 3 \ldots\ldots (2n) \cdot (p-1)(p-2) \ldots\ldots (p-2n)$.

Thus $x^2 \equiv 1 \cdot 2 \cdot 3 \ldots\ldots (2n) \cdot (p-2n) \ldots\ldots (p-2)(p-1)$,

where we have changed the order of the last $2n$ factors.

Now $2n = (p-1)/2$, thus $p - 2n = p - (p-1)/2 = (p+1)/2$,

$$p - (2n - 1) = (p+3)/2, \text{ and so on.}$$

Thus $x^2 \equiv 1 \cdot 2 \cdot 3 \ldots (p-1)/2 \cdot (p+1)/2 \cdot (p+3)/2 \ldots\ldots (p-1)$.

However the right hand side is just $(p-1)!$

Hence $x^2 \equiv (p-1)! \equiv -1$ (modulo p) by Wilson's theorem.

We require one more result before we can prove the main theorem due to Fermat mentioned earlier.

THEOREM 7.4.5 Let p be a prime integer. Let m be an integer co-prime (or relatively prime) to p, that is $(m, p) = 1$. Let $x^2 + y^2 = m\,p$ for certain integers x and y. Then $p = a^2 + b^2$, where a and b are integers.

PROOF

Since $\mathbf{Z} \subset \mathbf{Z}[i]$, p is also an element of $\mathbf{Z}[i]$. Suppose p is a prime in $\mathbf{Z}[i]$.

Now $\quad m\,p = x^2 + y^2 = (x + yi)(x - yi).$

Thus $\quad p \mid (x + yi)(x - yi)$ in the **Euclidean domain Z** $[i]$.

By theorem 7.2.5, $p \mid (x + yi)$ or $p \mid (x - yi)$. If $p \mid (x + yi)$, then $x + yi =$ $= p(c + di)$ for some $c + di \in \mathbf{Z}\,[i]$. Thus $x = p\,c$ and $y = p\,d$.

Hence p divides x and p divides y in **Z**.

Thus $\quad p^2 \mid x^2$ and $p^2 \mid y^2$ in **Z**.

Hence $\quad p^2 \mid (x^2 + y^2) = m\,p$ in **Z**.

But this means that $p \mid m$ in **Z**, which contradicts $(m, p) = 1$.

We get a similar contradiction if $p \mid (x - yi)$.

From these contradictions we deduce that p cannot be a prime in $\mathbf{Z}\,[i]$. This means that we can factorise p in $\mathbf{Z}\,[i]$ and write $p = (a + bi)(e + fi)$, where neither $a + bi$ nor $e + fi$ is a unit in $\mathbf{Z}\,[i]$.

The units of $\mathbf{Z}\,[i]$ are $\pm 1, \pm i$ (exercise 7.4).

Thus $\qquad$ ** $a^2 \geqq 1,\ b^2 \geqq 1,\ e^2 \geqq 1,\ f^2 \geqq 1.$

Since p is real, changing i into $-i$ does not affect $p = (a + bi)(e + fi)$. Thus we may also write $p = (a - bi)(e - fi)$.

Hence

$$p^2 = (a + bi)(e + fi)(a - bi)(e - fi)$$

$$= (a^2 + b^2)(e^2 + f^2).$$

This means that $(a^2 + b^2)$ divides p^2 in **Z**.

Hence $\qquad (a^2 + b^2) = 1, p,$ or p^2.

By ** above, we have $a^2 + b^2 \neq 1$ and $e^2 + f^2 \neq 1$.

Thus $\quad a^2 + b^2 \neq 1$ and $a^2 + b^2 \neq p^2$.

We conclude that $a^2 + b^2 = p$, as required.

Finally we prove

THEOREM 7.4.6 (Fermat) Let p be a prime integer of the form $4n + 1$, where

n is a positive integer. Then $p = a^2 + b^2$ for some integers a and b.

PROOF

By theorem 7.4.4 there exists an integer x so that $x^2 \equiv -1$ (modulo p). This means that $p \mid (x^2 + 1)$, so that $x^2 + 1 = m\,p$ for some integer m.

Writing the above congruence in the form:

$$\bar{x}^2 = (\overline{-1}),\ \bar{x} \in \mathbf{Z}_p = \{\bar{0}, \bar{1}, \ldots\ldots, (\overline{p-1})\},$$

we see that we may assume $0 \leqq x \leqq p - 1$. Moreover, since $(\overline{p-x})^2 = (\overline{-1})$ we could equally well take $p - x$ in place of x.

Noting that if $x > p/2$ then $p - x < p/2$, this means that we can further assume that $0 \leqq x < p/2$.

Thus $m\,p = x^2 + 1 < p^2/4 + 1 < p^2$.

Hence $m < p$. In particular $p \nmid m$, so that $(m, p) = 1$.

By theorem 7.4.5 $\quad p = a^2 + b^2$ for some integers a and b.

EXAMPLE 7.4.1 The first few primes of form $4n + 1$ are: 5, 13, 17, 29, 37, 41, 53, 61, and 73.

They may be written in the form $a^2 + b^2$ for suitable integers a and b as follows:

$$5 = 1^2 + 2^2, \qquad 13 = 2^2 + 3^2, \qquad 17 = 1^2 + 4^2,$$

$$29 = 2^2 + 5^2, \qquad 37 = 1^2 + 6^2, \qquad 41 = 4^2 + 5^2,$$

$$53 = 2^2 + 7^2, \qquad 61 = 5^2 + 6^2, \qquad 73 = 3^2 + 8^2.$$

EXERCISES

1. Find the greatest common divisor (G.C.D) of the numbers 741 and 1079 in the Euclidean domain **Z**.

Express the G.C.D in the form $A(741) + B(1079)$, for some $A, B \in \mathbf{Z}$.

2. Find the G.C.D $d(x)$ of $f(x)$ and $g(x)$, for the following pairs of polynomials in $\mathbf{Q}[x]$. Also find polynomials $F(x)$ and $G(x)$ in $\mathbf{Q}[x]$ so that:

$$d(x) = F(x)\,f(x) + G(x)\,g(x).$$

(a) $f(x) = x^3 - x^2 + x - 1, \qquad g(x) = x^4 - x^3 + x - 1$

(b) $f(x) = x^3 + x + 1, \qquad g(x) = x^2 - x + 1$

(c) $f(x) = x^4 + 1, \qquad g(x) = x^4 + x^2 + x + 1$

(d) $f(x) = x^4 - x^3 + 3x^2 - 2x + 2, \quad g(x) = x^4 + x^2 - 2$

(e) $f(x) = x^5 + x + 1, \quad g(x) = x^4 + x^2 + 2.$

3. Let $\mathbf{Z}[i]$ be the ring of Gaussian integers $\{a + bi : a, b \in \mathbf{Z}\}$. Define the function $\delta : \mathbf{Z}[i] - \{0\} \to \{0,1,2,3, \ldots\ldots\}$ by $\delta(a + bi) = a^2 + b^2$.

Then, as proved in chapter 7, $\mathbf{Z}[i]$ is a Euclidean domain.

Find a G.C.D of $1 + 5i$ and $4 + 2i$ and express in the form : $A(1 + 5i) + B(4 + 2i)$, for some $A, B \in \mathbf{Z}[i]$.

4. With the notation of exercise 3, show that with $x \neq 0, y \neq 0$:

(a) if $x \mid y$ in $\mathbf{Z}[i]$, then $\delta(x) \mid \delta(y)$ in $\{0,1,2,3,\ldots\ldots\}$.

(b) x is a unit in $\mathbf{Z}[i]$ if and only if $\delta(x) = 1$.

Prove that the units of $\mathbf{Z}[i]$ are $1, -1, i, -i$. What type of group do the units form?

5. Let R be a commutative ring with identity. Show that R is a field if and only if the only ideals of R are the zero ideal $\{0\}$ and the whole ring R.

6. Let $\mathbf{Z}[i\sqrt{5}]$ be the ring $\{a + bi\sqrt{5} : a, b \in \mathbf{Z}\}$.

Define the function $\delta : \mathbf{Z}[i\sqrt{5}] - \{0\} \to \{0,1,2,\ldots.\}$ by $\delta(a + bi\sqrt{5}) = a^2 + 5b^2$.

Let $x, y \in \mathbf{Z}[i\sqrt{5}]$. Prove

(a) $\delta(xy) = \delta(x)\,\delta(y);\ x, y \neq 0,$

(b) $\delta(x) = 1$ if and only if x is a unit,

(c) if $x \neq 0$ and $y \neq 0$ and y is not a unit, then $\delta(x) < \delta(xy)$,

(d) $2, 3, 1 + i\sqrt{5},\ 1 - i\sqrt{5}$ are primes,

(e) $\mathbf{Z}[i\sqrt{5}]$ is **not** a Euclidean domain.

Solutions to the Exercises

Exercises 1

1. $\sim(p \vee r)$ is true, hence p is false and r is false. $p \vee q$ is true, hence p or q is true. Hence q is true. Also $p \vee s$ is true. But p is false. Hence s is true. Thus p is false, q true, r false, s true. Hence, of the listed propositions, q and s are the necessarily true ones.

2. $p^2 - 1 = (p-1)(p+1)$. p odd so $p-1$ and $p+1$ consecutive even, say $2n$ and $2n+2$. If $n = 2r+1$ odd, then $2n = 4r+2$ and $2n+2 = 4r+4$. If $n=2r$ even, then $2n = 4r$ and $2n+2 = 4r+2$. Thus always 4 divides one, and 2 divides the other. Hence always 8 divides $(p-1)(p+1) = p^2 - 1$.

3. Tautologies are (*a*), (*c*), (*e*), (*f*), (*g*). The negation of (*d*) is a tautology.

4. Let $\sqrt{2} = p/q$; $p, q \in \mathbf{Z}^+$. Let $(p, q) = 1$. Then $2 = p^2/q^2$. Thus $p^2 = 2q^2$. Then $2 \mid p^2$. Hence p is even, say $p = 2r$. Then $2q^2 = 4r^2$. Thus $q^2 = 2r^2$. Hence $2 \mid q^2$. Thus $q = 2s$. But then $(p, q) \neq 1$. Contradiction. Hence $\sqrt{2}$ is not rational.

5d. (1) $(p \to ((p \to p) \to p)) \to ((p \to (p \to p)) \to (p \to p))$ by (b).

(2) $p \to ((p \to p) \to p)$ by (a).

(3) $(p \to (p \to p)) \to (p \to p)$ from (1) and (2) by modus ponens.

(4) $p \to (p \to p)$ by (a).

(5) $p \to p$ from (3) and (4) by modus ponens.

(e) (1) $(\sim p \to \sim p) \to ((\sim p \to p) \to p)$ by (c).

(2) $\sim p \to \sim p$ by (d), just proved.

(3) $(\sim p \to p) \to p$ from (1) and (2) by modus ponens.

6. Assume x odd **and** x^2 even ($x \in \mathbf{Z}$). If x is odd, then $x = 2k + 1$ ($k \in \mathbf{Z}$). Thus $x^2 = (2k + 1)^2 = 4k^2 + 4k + 1 = 2(2k^2 + 2k) + 1$, which is odd. But x^2 is even, by hypothesis. Hence the hypothesis is false, and the given statement is true.

7. Assume $x \neq 0$. Then $|x| > 0$. Thus there exists $y > 0$ such that $|x| \geq y$. ($y = |x|$ will do).

8. Assume lines M and N intersect at point p. Let acute angle between lines M and N at p be $\theta > 0$. Let alternate angles be α and β.

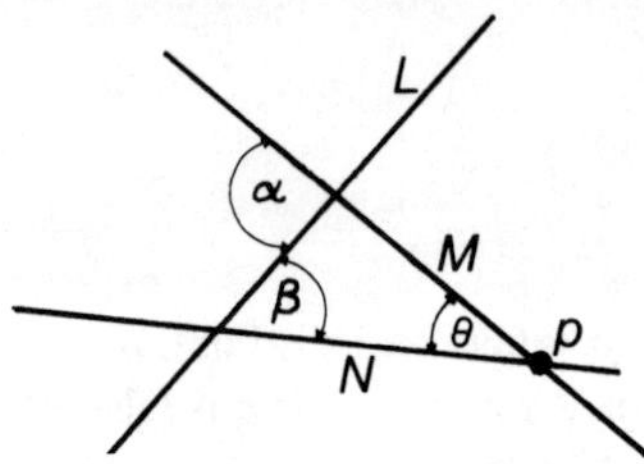

Then the sum of the interior angles of the triangle formed by the lines L, M, and N is:

$$\beta + (180^\circ - \alpha) + \theta \quad = \quad \beta + (180^\circ - \beta) + \theta =$$

$$= 180^\circ + \theta > 180^\circ; \qquad \text{since } \alpha = \beta, \text{ given.}$$

But this contradicts the assumption that the sum of the interior angles of a triangle is 180°. We deduce that lines M and N do not intersect.

9. (a) Domain of interpretation is the set of real numbers $\mathbf{R}$. $\exists x. (x^2 = 2)$. We could express the domain of interpretation by writing just $x \in \mathbf{R}$. Then: $\exists x. (x^2 = 2),\ x \in \mathbf{R}$.

(b) $\forall n\ (n^2 \geqq 0),\ n \in \mathbf{Z}$.

(c) $\forall x\ \forall y\ (\ ((x \in \mathbf{Q}) \wedge (y \in \mathbf{Q})) \rightarrow\ (x + y \in \mathbf{Q})\ )$,

where $\mathbf{Q}$ denotes the set of rational numbers.

10. $x^2 + 1 = 0$ has no real solutions, so certainly has no rational solutions.

11. The three statements (a), (b), (c) are all false. The statement (d) is true.

Exercises 2

1. 9 in S, 7 in T, 3 in both S and T.

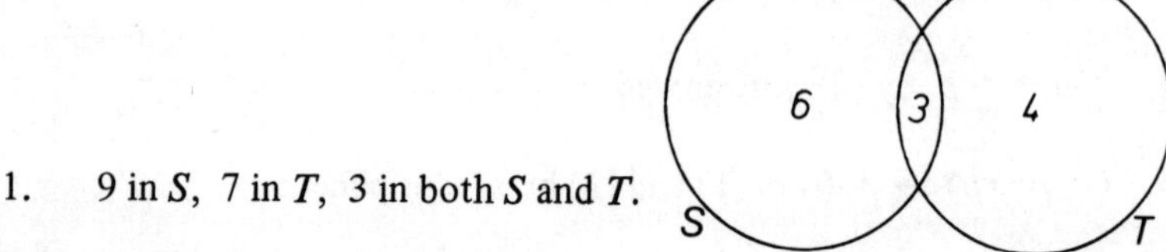

Hence 6 in S not in T, 4 in T not in S. Thus 13 in $S \cup T$. Thus $|S \cup T| = 13$.

2. Let $|S \cap T| = z$, $|S| = x$, $|T| = y$. Then

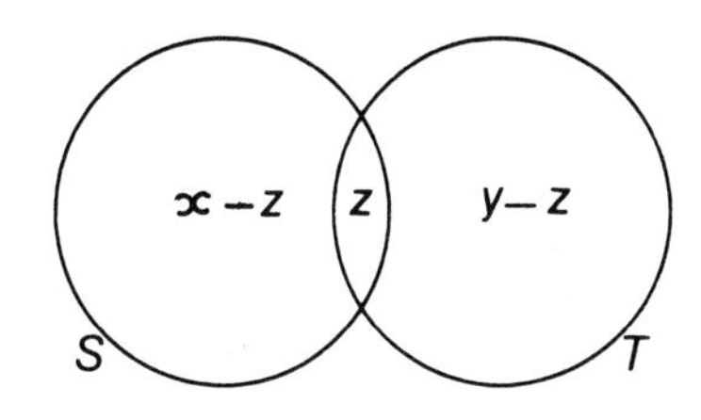

$$\text{Thus } |S \cup T| = (x - z) + (y - z) + z$$
$$= x + y - z.$$

Hence $|S \cup T| = |S| + |T| - |S \cap T|$.

3.

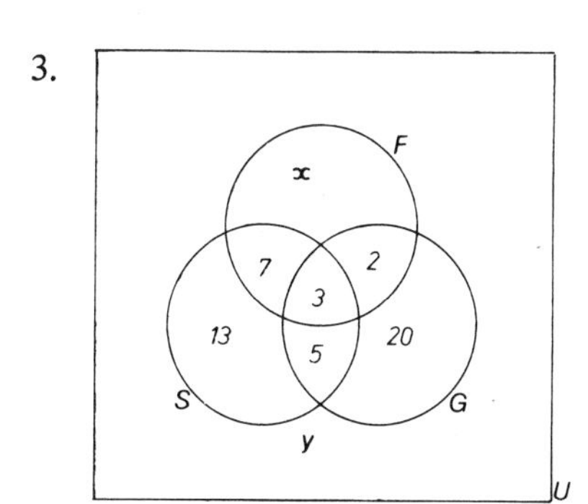

Label diagram as shown.
x study French alone.
y study no language.
Then $x = 42 - 12 = 30$.
$y = 100 - (13 + 7 + 3 + 5 + 2 + 20 + 30)$
$= 20$.

4. $f \circ g = \{(1,3), (2,2), (3,1)\}$
$g \circ f = \{(1,1), (2,3), (3,2)\}$, $f \circ f = \{(1,3), (2,1), (3,2)\}$
$f \circ g \circ f = \{(1,2), (2,1), (3,3)\}$, $f^{-1} = \{(1,3), (2,1), (3,2)\}$
By inspection $f \circ g \circ f = g$ and $f \circ f = f^{-1}$.

5. (a) Take any $x \in C$. Then $g(b) = x$, some $b \in \mathbf{B}$, and $f(a) = b$, some $a \in A$. Then $(g \circ f)(a) = g(b) = x$. Thus $g \circ f$ is surjective.

(b) Let $(g \circ f)(a_1) = (g \circ f)(a_2)$. Then $g(f(a_1)) = g(f(a_2))$.
g is injective, hence $f(a_1) = f(a_2)$.
Now f is injective, hence $a_1 = a_2$.
Thus $g \circ f$ is injective.

6. We have to show that $f(A_1) \cap f(A_2) = \phi$ and $f(A_1) \cup f(A_2) = f(A)$. Since $A_1 \subset A$ and $A_2 \subset A$ we have $f(A_1) \subset f(A)$ and $f(A_2) \subset f(A)$. Thus $f(A_1) \cup f(A_2) \subset f(A)$. Let $x \in f(A)$. Then $x = f(a)$ for some $a \in A$. But $A = A_1 \cup A_2$, thus $a \in A_1$ or $a \in A_2$. Suppose $a \in A_1$. Then $f(a) \in f(A_1)$. If $a \in A_2$, then $f(a) \in f(A_2)$. Thus, in any case, $f(a) \in f(A_1) \cup f(A_2)$. Hence $f(A) \subset f(A_1) \cup f(A_2)$. Altogether we get: $f(A_1) \cup f(A_2) = f(A)$. Now suppose $x \in f(A_1) \cap f(A_2)$. Then $x = f(a_1)$, $a_1 \in A_1$, and $x = f(a_2)$, $a_2 \in A_2$. Hence $f(a_1) = f(a_2)$. f is injective so $a_1 = a_2$. Then $a_1 = a_2 \in A_1 \cap A_2 = \phi$. Contradiction. Hence $f(A_1) \cap f(A_2) = \phi$.

7. $f\colon \mathbf{N} \to S$, where $f(n) = n + k$ and $0 \in \mathbf{N}$, is a bijection. By definition, S has cardinality $\aleph_0$.

8. Define $f\colon T \longrightarrow \mathbf{R}$, where $T = \{x \in \mathbf{R} : -1 < x < 1\}$, by $f(x) = \tan(\pi x)/2$. Then f is a bijection. By definition $|T| = |\mathbf{R}| = c$.

9. Suppose S and $\mathcal{P}(S)$ have same cardinality. Let $f\colon S \to \mathcal{P}(S)$ be the corresponding bijection. Let $T = \{x \in S : x \notin f(x)\} \subset S$. For some $s \in S, f(s) = T$. Then $s \in T \to s \notin T$ and $s \notin T \to s \in T$. Contradiction. Hence no bijection $f\colon S \to \mathcal{P}(S)$ exists. (If $S = \phi$, then $\mathcal{P}(S) = \{\phi\}$. S has no members. $\mathcal{P}(S)$ has one member).

10. If $a \neq 1$, then a and a have common factor $a \neq 1$. If $a = 1$, then $1\ R\ 1$ given. Thus $a\ R\ a$ and R is reflexive. Symmetry is immediate. $12\ R\ 8$ and $8\ R\ 12$ but $8 \neq 12$. Thus R is not anti-symmetric. $9\ R\ 12$ and $12\ R\ 8$ but 9 is not R 8. Thus R is not transitive.

11. Let r, s and t denote reflexive, symmetric and transitive respectively. Then:

$\{(1,1), (2,2), (3,3), (4,4), (5,5), (6,6), (1,2), (2,1), (5,6), (6,5)\}$ r,s,t

$\{(1,1), (2,2), (3,3), (4,4), (5,5), (6,6), (1,2), (2,1), (2,3), (3,2)\}$ r,s

$\{(1,1), (2,2), (3,3), (4,4), (5,5), (6,6), (1,2), (2,3), (1,3)\}$ r,t

$\{(1,1), (2,2), (3,3), (4,4), (5,5), (6,6), (1,2), (2,3)\}$ r

$\{(1,2), (2,1), (1,1), (2,2)\}$ s,t

$\{(1,2), (2,1)\}$ s, $\{(1,2), (2,3), (1,3)\}$ t

$\{(1,2), (2,3)\}$ not s, not t, not r.

A directed graph describing the fifth of the above is:

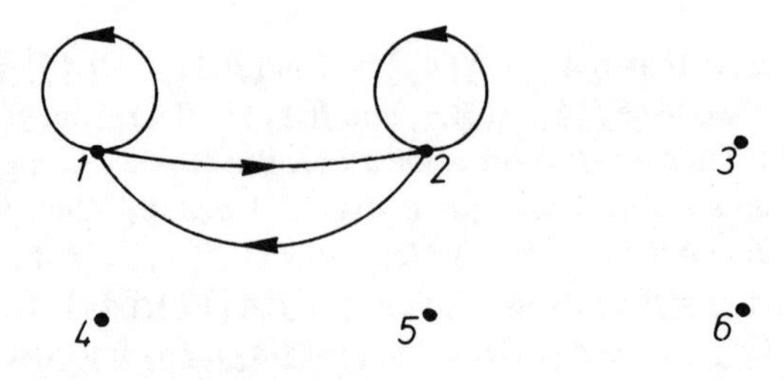

Exercises 3

1. (a) $6 \mid (a - a)$, hence $a \equiv a$ (reflexive). If $6 \mid (a - b)$, then $6 \mid (b - a)$. Thus if $a \equiv b$, then $b \equiv a$ (symmetric). If $a \equiv b$, and $b \equiv c$, then $a = b + 6r$, $b = c + 6s$. Thus $a = (c + 6s) + 6r$. Hence $a - c = 6(s + r)$. Thus $6 \mid (a - c)$. Hence $a \equiv c$ (transitive). Thus $\equiv$ is an equivalence relation.

(b) Suppose $\overline{a} = \overline{a'}$ and $\overline{b} = \overline{b'}$.

Then $a = a' + 6r$ and $b = b' + 6s$.

Thus $a + b = a' + b' + 6(r + s)$.

Thus $\overline{(a + b)} = \overline{(a' + b')}$.

Similarly for $\overline{a} \cdot \overline{b} = \overline{a'} \cdot \overline{b'}$.

(c) $\overline{a}\,(\overline{b} \cdot \overline{c}) = \overline{a}\,(\overline{bc}) = \overline{a(bc)} = \overline{(ab)c} = (\overline{a} \cdot \overline{b})\,\overline{c}$.

Thus associative law holds in $\mathbf{Z}_6$. Similarly for the other ring axioms. In a sense they are inherited from $\mathbf{Z}$. $\overline{0}$ is the zero and $\overline{1}$ is the identity in $\mathbf{Z}_6$.

(d) $\overline{2} \cdot \overline{3} = \overline{6} = \overline{0}$ but $\overline{2} \neq \overline{0}$ and $\overline{3} \neq \overline{0}$. Thus $\mathbf{Z}_6$ is **not** an integral domain.

(e) Suppose $\mathbf{Z}_6$ were a field. Then $\overline{2} \neq \overline{0}$ would have an inverse $\overline{a}$, say.

Then $\overline{2} \cdot \overline{a} = \overline{1}$. Hence $\overline{3} \cdot (\overline{2}\,\overline{a}) = \overline{3} \cdot \overline{1} = \overline{3}$.

However $\overline{3}(\overline{2}\,\overline{a}) = (\overline{3}\,\overline{2})\,\overline{a} = \overline{6}\,\overline{a} = \overline{0}\,\overline{a} = \overline{0}$. Thus $\overline{3} = \overline{0}$.

Contradiction. Hence $\overline{2}$ does not have an inverse. Thus $\mathbf{Z}_6$ is not a field.

2. Let $\mathbf{Z}_p = [\overline{0}, \overline{1}, \overline{2}, \ldots\ldots \overline{(p - 1)}]$ with p prime.

Let $\overline{a} \neq \overline{0}$. Then $(a, p) = 1$.

Thus $Aa + Pp = 1$. Thus $\overline{A}\overline{a} + \overline{P}\overline{p} = \overline{1}$. Hence $\overline{A}\overline{a} = \overline{1}$.

Thus $\overline{a}$ has an inverse $\overline{A}$. Hence $\mathbf{Z}_p$ is a field.

See chapter 7 for the theory which gives $Aa + Pp = 1$ etc. Since 5 is small direct enumeration is feasible in this case. Thus

$$\mathbf{Z}_5 = \{\overline{0}, \overline{1}, \overline{2}, \overline{3}, \overline{4}\}. \quad (\overline{1})^{-1} = \overline{1}, \quad (\overline{2})^{-1} = \overline{3}, \quad (\overline{3})^{-1} = \overline{2}, \quad (\overline{4})^{-1} = \overline{4}.$$

3. Let $P(n)$ denote: $1^2 + 2^2 + \ldots\ldots + n^2 = n(n + 1)(2n + 1)/6$.

Then $P(1)$: $1 = 1(2)(3)/6$ is true. Now prove implication $P(n) \rightarrow P(n + 1)$.

From $P(n)$ we have:

$$1^2 + 2^2 + \ldots\ldots + n^2 + (n+1)^2 = n(n+1)(2n+1)/6 + (n+1)^2$$
$$= (n+1)((n+1)+1)(2(n+1)+1)/6$$

which is $P(n+1)$. From $P(1)$ and $P(n) \to P(n+1)$ true, induction gives $P(n)$ true for all positive integers n.

4. As above, using $1 + 2 + 3 + \ldots\ldots + n = n(n+1)/2$.

$$P(n) \to \{1^3 + 2^3 + \ldots\ldots + n^3 + (n+1)^3\} = (1 + 2 + \ldots + n)^2 + (n+1)^3 =$$

$$= \{n(n+1)/2\}^2 + (n+1)^3 = \{(n+1)(n+2)/2\}^2 =$$

$$= (1 + 2 + \ldots\ldots + n + (n+1))^2.$$

Thus $P(n) \to P(n+1)$.

5. (a) $(2+i)z + i = 3 \to z = (3-i)/(2+i) \to$

$\to \quad z = (3-i)(2-i)/(2+i)(2-i) = (3-i)(2-i)/5 = (1-i)$.

(b) $(z-1)/(z-i) = 2/3 \to 3(z-1) = 2(z-i) \to$

$\to \quad z = 3 - 2i$.

(c) $z^2 - (3+i)z + 4 + 3i = 0 \to z = [(3+i) \pm \sqrt{-8-6i}]/2$.

Put $\sqrt{-8-6i} = x + iy$ (x,y real).

Then $-8 - 6i = (x+iy)^2 \to x^2 - y^2 = -8,\ xy = -3 \to x = 1,$

$y = -3 \to z = (2-i)$ or $(1+2i)$.

(d) $z^4 - 2z^2 + 4 = 0 \to (z^2)^2 - 2(z^2) + 4 = 0 \to$

$\to \quad z^2 = (2 \pm \sqrt{4-16})/2 \to z^2 = 1 \pm \sqrt{3}\,i$.

Put $z = x + iy$. Then $x^2 - y^2 = 1,\ 2xy = \pm\sqrt{3}$.

Hence $\quad z = \pm(\sqrt{3} \pm i)/\sqrt{2}$; with 4 possible results.

6. $|(1+2i)^{12}/(1-2i)^9| = |1+2i|^{12}/|1-2i|^9 = (\sqrt{5})^{12}/(\sqrt{5})^9 =$

$= (\sqrt{5})^3 = 5\sqrt{5}$.

7. Immediate from a diagram.

Let $A = 1 + 2i,\ B = 2 + 3i,\ C = 3 + 2i,\ D = 2 + i$.

In coordinates $A = (1,2)$, $B = (2,3)$, $C = (3,2)$, $D = (2,1)$.

Line AC bisects line BD at right angles. Thus $ABCD$ is a square.

8. 'Sum of distances from point to foci is constant' defines an ellipse. Hence the region is the interior and boundary of an ellipse with foci at 1 and i and major axis 4. Alternatively, write $z = x + iy$ and convert to a problem in co-ordinate geometry on the Argand plane.

9. $(z - 2i)/(2z - 1) = ai$, a real.

Let $z = x + iy$. Then $x + iy - 2i = ai(2x + 2iy - 1)$.

Thus $x = -2ay$ and $y - 2 = (2x - 1)a$.

Eliminate a. Then $x = -2y(y - 2)/(2x - 1)$.

Thus $(2x - 1)x + 2y(y - 2) = 0 \rightarrow 2(x^2 + y^2) - x - 4y = 0$

$\rightarrow \quad x^2 + y^2 - x/2 - 2y = 0 \rightarrow (x - 1/4)^2 + (y - 1)^2 = 17/16$.

Circle centre $(1/4, 1) = 1/4 + i$, radius $\sqrt{17}/4$.

10. $(z - 6i)/(z + 8) = a$ (real).

Put $z = x + iy$. Then $x + i(y - 6) = a(x + 8) + iya$.

Thus $x = a(x + 8)$ and $y - 6 = ya$. Eliminate a: $y - 6 = yx/(x + 8) \rightarrow$

$\rightarrow \quad (y - 6)(x + 8) = xy \rightarrow 8y - 6x - 48 = 0$.

Thus $y = 3x/4 + 6$, which is a straight line.

11.

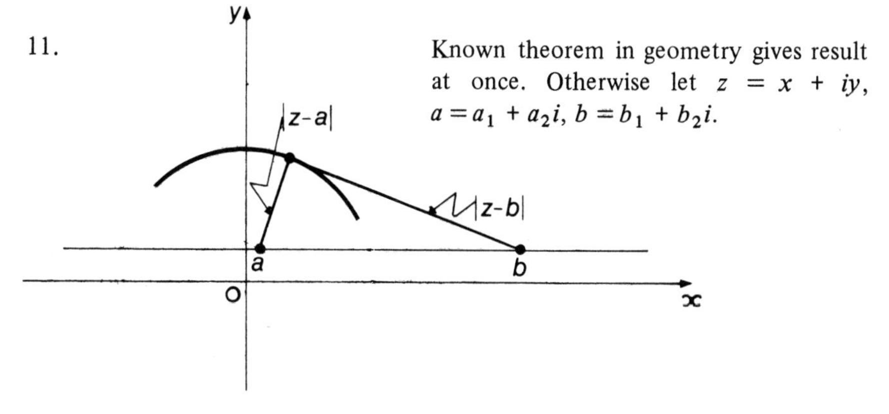

Known theorem in geometry gives result at once. Otherwise let $z = x + iy$, $a = a_1 + a_2 i$, $b = b_1 + b_2 i$.

Then $|z - a|^2 = \lambda^2 |z - b|^2$ gives:

$$\lambda < 1 \;.\; (1 - \lambda^2)(x^2 + y^2) + (2b_1\lambda^2 - 2a_1)x + (2b_2\lambda^2 - 2a_2)y + a_1^2$$

$$+ a_2^2 - \lambda^2(b_1^2 + b_2^2) = 0.$$

This is a circle. With $|z - a| < |z - b|$, the circle 'curves' towards a, so that a is inside the circle. The case when $\lambda > 1$ follows from the above if we write $|(z - b)/(z - a)| = 1/\lambda < 1$, where the roles of a and b are interchanged and $\lambda' = 1/\lambda < 1$.

12. $(1 + \sin\theta - i\cos\theta)(\sin\theta + i\cos\theta) =$

$$= \sin\theta + \sin^2\theta - i\sin\theta\cos\theta + i\cos\theta + i\sin\theta\cos\theta + \cos^2\theta =$$

$$= 1 + \sin\theta + i\cos\theta.$$

Thus $(1 + \sin\theta + i\cos\theta)/(1 + \sin\theta - i\cos\theta) = \sin\theta + i\cos\theta.$

Put $\theta = \pi/5$.

Then $[(1 + \sin\pi/5 + i\cos\pi/5)/(1 + \sin\pi/5 - i\cos\pi/5)]^5 =$

$$= (\sin\pi/5 + i\cos\pi/5)^5 = [\cos(\pi/2 - \pi/5) + i\sin(\pi/2 - \pi/5)]^5 =$$

$$= \cos(5\pi/2 - \pi) + i\sin(5\pi/2 - \pi) = \cos 3\pi/2 + i\sin 3\pi/2 =$$

$= 0 + i(-1) = -i.$ Required result follows.

13. $1^{1/6} = (\cos 2m\pi + i\sin 2m\pi)^{1/6} =$

$$= \cos 2m\pi/6 + i\sin 2m\pi/6.$$

Put $m = 0,1,2,3,4,5$ in turn to get the six 1/6th roots of unity. These are:

$$\pm 1,\ \tfrac{1}{2} \pm i\sqrt{3}/2,\ -\tfrac{1}{2} \pm i\sqrt{3}/2.$$

14. $2 + 2i = \sqrt{8}\left(\frac{1}{\sqrt{2}} + \frac{i}{\sqrt{2}}\right) = \sqrt{8}(\cos\pi/4 + i\sin\pi/4) =$

$= \sqrt{8}(\cos(2m\pi + \pi/4) + i\sin(2m\pi + \pi/4))$. By de Moivre,

$(2 + 2i)^{1/3} = (8)^{1/6}(\cos(8m + 1)\pi/12 + i\sin(8m + 1)\pi/12)$, for $m = 0,1,2$.

The 3 possible values of

$(2 + 2i)^{1/3}$ are: $\sqrt{2}(\cos\pi/12 + i\sin\pi/12)$, $-1 + i$, and

$-\sqrt{2}(\cos 5\pi/12 + i\sin 5\pi/12)$.

15. $z = \cos\theta + i\sin\theta$. $\cos\theta = -½$.

Thus $\theta = {}^{2\pi}/_3$, or ${}^{4\pi}/_3$. Hence $\sin\theta = \sqrt{3}/2$, or $-\sqrt{3}/2$.

Then $z = -½ + i\sqrt{3}/2$ or $z = -½ - i\sqrt{3}/2$.

Thus $z^3 = 1$ in both cases. Alternatively, write:

$$z = \cos{}^{2\pi}/_3 + i\sin{}^{2\pi}/_3 \text{ or } z = \cos{}^{4\pi}/_3 + i\sin{}^{4\pi}/_3.$$

Then de Moivre gives:

$$z^3 = \cos 2\pi + i\sin 2\pi = 1, \text{ or } z^3 = \cos 4\pi + i\sin 4\pi = 1.$$

16.

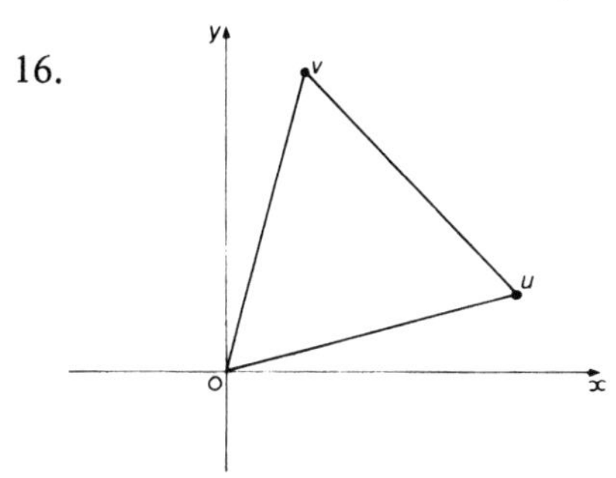

Suppose 0, u, v is an equilateral triangle on the Argand diagram. Then $u\,e^{i\pi/3} = v$. (Multiplication by $e^{i\theta}$ does not alter the modulus, but increases the argument by θ).

Thus $u^2 + v^2 =$

$$= u^2 + (u\,e^{i\pi/3})^2 = u^2 + u^2\,e^{i2\pi/3} = u^2\,(1 + \cos{}^{2\pi}/_3 + i\sin{}^{2\pi}/_3) =$$

$$= u^2\,(1 + (-½) + i\sqrt{3}/2) = u^2\,(½ + i\sqrt{3}/2) = u^2\,e^{i\pi/3} = u \cdot v.$$

Conversely, suppose $u^2 + v^2 = uv$.

Let $v = k\,u\,e^{i\theta}$, where k is real, and $\arg v = \arg u + \theta$.

This amounts to rotating the line $0u$ to coincidence with the line from 0 to v, and then 'stretching' or 'contracting' until we get coincidence with $0v$. From $u^2 + v^2 = uv$ we get:

$$u^2 + k^2\,u^2\,e^{i2\theta} = k\,u^2\,e^{i\theta}.$$

Thus $1 + k^2\,e^{i\theta} = k\,e^{i\theta}$ or $(k\,e^{i\theta})^2 - (k\,e^{i\theta}) + 1 = 0$.

This gives: $k\,e^{i\theta} = e^{i\pi/3}$ or $e^{i5\pi/3}$.

Thus $k = 1$ and $\theta = 60$ or 300 degrees. In either case the triangle 0, u, v is equilateral.

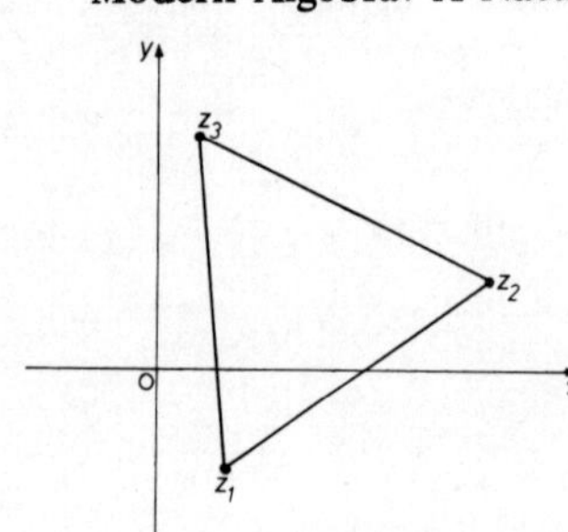

Put $u = z_2 - z_1$ and $v = z_3 - z_1$. Then, using the above result, z_1, z_2, z_3 are the vertices of an equilateral triangle if and only if

$$(z_2 - z_1)^2 + (z_3 - z_1)^2 = (z_2 - z_1)(z_3 - z_1)$$

This reduces to: $\quad z_1^2 + z_2^2 + z_3^2 = z_1 z_2 + z_1 z_3 + z_2 z_3.$

17. $z^5 = -1 = \cos(2m + 1)\pi + i \sin(2m + 1)\pi$. By de Moivre

$$z = \cos(2m + 1)\pi/5 + i \sin(2m + 1)\pi/5; \quad m = 0,1,2,3,4.$$

Roots of $z^5 + 1 = 0$ are: $\cos {}^{\pi}/_5 \pm i \sin {}^{\pi}/_5, \cos {}^{3\pi}/_5 \pm i \sin {}^{3\pi}/_5, -1.$

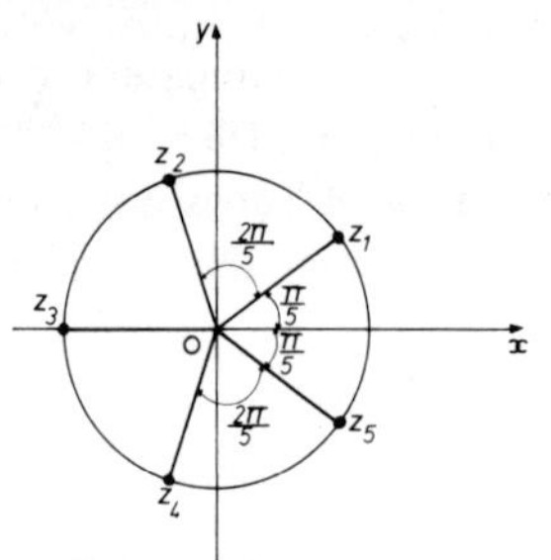

Roots are marked z_1, z_2, z_3, z_4, z_5 on the Argand diagram. The circle is the unit circle, centre 0. Angle between consecutive roots is: ${}^{2\pi}/_5$ radians = 72 degrees.

18. From exercise 17 above, roots of $z^5 + 1 = 0$ are z_1, z_2, z_3, z_4, z_5. Thus

$$z^5 + 1 \equiv (z - z_1)(z - z_2)(z - z_3)(z - z_4)(z - z_5)$$

$$\equiv (z + 1)(z^2 - (z_2 + z_4)z + z_2 z_4)(z^2 - (z_1 + z_5)z + z_1 z_5) \equiv$$

$$\equiv (z + 1)(z^2 - 2z \cos {}^{\pi}/_5 + 1)(z^2 - 2z \cos {}^{3\pi}/_5 + 1)$$

Put $\quad z = i.\quad$ Then $\quad i^5 + 1 = (i + 1)(i^2 - 2i \cos {}^{\pi}/_5 + 1)(i^2 - 2i \cos {}^{3\pi}/_5 + 1)$

Now $\quad i^5 + 1 = i + 1.\quad$ Thus $\quad 1 = -4 \cos {}^{\pi}/_5 \cos {}^{3\pi}/_5.$

However $\cos {}^{3\pi}/_5 = \sin({}^{\pi}/_2 - {}^{3\pi}/_5) = \sin(-{}^{\pi}/_{10}) = -\sin {}^{\pi}/_{10}.$

Hence $\quad 1 = 4 \sin {}^{\pi}/_{10} \cos {}^{3\pi}/_5.$

19. $(z - 1)^6 + (z + 1)^6 = 0.\quad$ Put $\quad w = (z + 1)/(z - 1).$

Then $w^6 = -1.\quad$ Thus $\quad w = (-1)^{1/6} = \cos(2m + 1)\pi/6 + i \sin(2m + 1)\pi/6$

for $\quad m = 0,1,2,3,4,5.$

Thus $w = \cos s\pi/6 \pm i \sin s\pi/6; \; s = 1,3,5.$

From $w = (z+1)/(z-1)$, we get $z = (w+1)/(w-1)$.

Put $w = \cos\theta + i\sin\theta$.

Then $z = (\cos\theta + i\sin\theta + 1)/(\cos\theta + i\sin\theta - 1) =$

$$= \frac{\cos^2\theta/2 - \sin^2\theta/2 + i\,2\sin\theta/2\cos\theta/2 + \cos^2\theta/2 + \sin^2\theta/2}{\cos^2\theta/2 - \sin^2\theta/2 + i\,2\sin\theta/2\cos\theta/2 - \cos^2\theta/2 - \sin^2\theta/2} =$$

$$= \frac{\cos^2\theta/2 + i\sin\theta/2\cos\theta/2}{-\sin^2\theta/2 + i\sin\theta/2\cos\theta/2} = \frac{\cot\theta/2\,(\cot\theta/2 + i)}{i\,(i + \cot\theta/2)} =$$

$$= (\cot\theta/2)/i = -i\cot\theta/2.$$

Put $\theta = \pi/6, 3\pi/6, 5\pi/6$ in turn; also take complex conjugates.

Then all roots of $(z-1)^6 + (z+1)^6 = 0$ are: $\pm i\cot{}^{\pi}/_{12}$, $\pm i$, $\pm i\cot{}^{5\pi}/_{12}$.

20. $\cos 4\theta + i\sin 4\theta = (\cos\theta + i\sin\theta)^4 =$

$$= \cos^4\theta + 4i\cos^3\theta\sin\theta - 6\cos^2\theta\sin^2\theta - 4i\cos\theta\sin^3\theta + \sin^4\theta$$

(by binomial theorem).

Thus $\cos 4\theta = \cos^4\theta - 6\cos^2\theta\sin^2\theta + \sin^4\theta =$

$$= \cos^4\theta - 6\cos^2\theta\,(1-\cos^2\theta) + (1-\cos^2\theta)^2 =$$

$$= 8\cos^4\theta - 8\cos^2\theta + 1.$$

21. Put $z = \cos\theta + i\sin\theta$.

Then $2\cos\theta = z + z^{-1}$. In general $z^n + z^{-n} = 2\cos n\theta$.

Thus $2^6\cos^6\theta = (z + z^{-1})^6 =$

$$= (z^6 + z^{-6}) + 6(z^4 + z^{-4}) + 15(z^2 + z^{-2}) + 20,$$

using binomial and collecting terms.

Hence $2^6\cos^6\theta = 2\cos 6\theta + 12\cos 4\theta + 30\cos 2\theta + 20.$

Thus $\cos^6\theta = (\cos 6\theta + 6\cos 4\theta + 15\cos 2\theta + 10)/32.$

Note. $\sin^6 \theta$ can be found using : $2i \sin \theta = z - z^{-1}$.

Exercises 4

1. (a) Not possible. (b) $\begin{pmatrix} 3 & 1 & -1 \\ 0 & 0 & 2 \\ -2 & 2 & 2 \end{pmatrix}$ (c) Not possible.

(d) $\begin{pmatrix} -1 & -1 & 1 \\ 0 & -2 & 2 \\ 0 & -2 & 0 \end{pmatrix}$ (e) Not possible. (f) Not possible.

(g) $\begin{pmatrix} 4 & 9 & -1 & -1 \\ 4 & 0 & 2 & 1 \\ 9 & -7 & 6 & 4 \end{pmatrix}$ (h) $\begin{pmatrix} 6 & -5 & 3 & 3 \\ 7 & -2 & 5 & 2 \end{pmatrix}$ (i) Not possible.

(j) Not possible. (k) $\begin{pmatrix} -1 & 3 & 1 \\ -2 & 1 & 1 \end{pmatrix}$ (l) $\begin{pmatrix} 3 & -1 & 1 \\ 0 & -1 & 2 \\ -2 & -2 & 5 \end{pmatrix}$

(m) $\begin{pmatrix} 2 & 1 & -1 \\ -2 & 3 & 2 \\ -3 & 1 & 2 \end{pmatrix}$ (n) Not possible. (o) $\begin{pmatrix} 1 & 2 & -1 & 0 \\ 0 & -10 & 0 & 3 \\ 1 & -7 & 2 & 2 \end{pmatrix}$

(p) Not possible. (q) $\begin{pmatrix} 2 & -2 & 2 \\ -1 & 2 & 0 \end{pmatrix}$ (r) $\begin{pmatrix} 4 & -6 \\ 0 & 1 \end{pmatrix}$

(s) $\begin{pmatrix} 8 & -14 \\ 0 & 1 \end{pmatrix}$ (t) $\begin{pmatrix} -2 & 2 & 4 \\ -3 & -1 & 5 \end{pmatrix}$ (u) $\begin{pmatrix} -2 & 2 & 4 \\ -3 & -1 & 5 \end{pmatrix}$

(v) $\begin{pmatrix} -4 & 6 & -2 \\ -1 & 2 & 0 \\ 10 & -10 & 10 \end{pmatrix}$ (w) $\begin{pmatrix} -4 & 6 & -2 \\ -1 & 2 & 0 \\ 10 & -10 & 10 \end{pmatrix}$ (x) $\begin{pmatrix} -1 & -3 & 5 \\ -1 & -2 & 4 \\ -5 & -5 & 15 \end{pmatrix}$

(y) $\begin{pmatrix} -1 & -3 & 5 \\ -1 & -2 & 4 \\ -5 & -5 & 15 \end{pmatrix}$ (z) $\begin{pmatrix} 10 & -9 \\ 2 & -2 \end{pmatrix}$ (z′) $\begin{pmatrix} 10 & -9 \\ 2 & -2 \end{pmatrix}$.

2. $$(A+B)^2 = (A+B)(A+B) = A^2 + AB + BA + B^2$$

$$(A-B)^2 = (A-B)(A-B) = A^2 - AB - BA + B^2.$$

In general $AB \neq BA$. Thus $AB + BA \neq 2AB, -AB - BA \neq -2AB$.

$$(A-B)(A+B) = A^2 - BA + AB - B^2.$$

In general $AB \neq BA$, so $-BA + AB \neq 0$.

Thus $(A-B)(A+B) \neq A^2 - B^2$, in general.

3. (a) Direct calculation.

(b) $ABA = A(BA) = AB = A$.

Also $ABA = (AB)A = A^2$. Thus $A^2 = A$.

$BAB = B(AB) = BA = B$.

Also $BAB = (BA)B = B^2$. Thus $B^2 = B$.

4. (a) $$A^2 = \begin{pmatrix} 0 & 0 & 0 \\ 3 & 3 & 9 \\ -1 & -1 & -3 \end{pmatrix} \neq 0.\ A^3 = 0.$$

(b) $M^2 = 0$. Thus $M^3 = M^4 \ldots . = M^n = 0$.

Hence $M(I \pm M)^n = M(I \pm nM) = M \pm nM^2 = M$, using binomial theorem.

5. (a) -6.

(b) $$\begin{vmatrix} 1 & 0 & 6 \\ 3 & 4 & 15 \\ 5 & 6 & 21 \end{vmatrix} = \begin{vmatrix} 1 & 0 & 6 \\ 0 & 4 & -3 \\ 0 & 6 & -9 \end{vmatrix} = \begin{vmatrix} 4 & -3 \\ 6 & -9 \end{vmatrix} = -18.$$

(c) -1.

6. $$\begin{vmatrix} 1 & a & b+c \\ 1 & b & c+a \\ 1 & c & a+b \end{vmatrix} = \text{Add col. 2 to col. 3} \begin{vmatrix} 1 & a & a+b+c \\ 1 & b & b+c+a \\ 1 & c & c+a+b \end{vmatrix} =$$

$$= (a+b+c) \begin{vmatrix} 1 & a & 1 \\ 1 & b & 1 \\ 1 & c & 1 \end{vmatrix} = 0, \text{ two columns are the same.}$$

7. $$\begin{vmatrix} 1 & a & a^2 \\ 1 & b & b^2 \\ 1 & c & c^2 \end{vmatrix} = \begin{vmatrix} 1 & a & a^2 \\ 0 & b-a & b^2-a^2 \\ 0 & c-a & c^2-a^2 \end{vmatrix} = (b-a)(c-a)$$

$$\begin{vmatrix} 1 & a+b \\ 1 & a+c \end{vmatrix} = (b-a)(c-a)(c-b).$$

8. $$\begin{vmatrix} bc & a^2 & a^2 \\ b^2 & ca & b^2 \\ c^2 & c^2 & ab \end{vmatrix} = \frac{1}{abc} \begin{vmatrix} abc & ba^2 & ca^2 \\ ab^2 & bca & cb^2 \\ ac^2 & bc^2 & cab \end{vmatrix} =$$

$$= \frac{1}{abc} . abc \begin{vmatrix} bc & ba & ca \\ ab & ca & cb \\ ac & bc & ab \end{vmatrix} .$$

9. $$A = \begin{pmatrix} 2 & 4 & 1 \\ 1 & 1 & 1 \\ 2 & 3 & 1 \end{pmatrix} . \text{ Matrix of cofactors } = \begin{pmatrix} -2 & 1 & 1 \\ -1 & 0 & 2 \\ 3 & -1 & -2 \end{pmatrix} = M.$$

$$\text{Adjugate of } A = M' = \begin{pmatrix} -2 & -1 & 3 \\ 1 & 0 & -1 \\ 1 & 2 & -2 \end{pmatrix} .$$

$$|A| = 2(-2) + 4(1) + 1(1) = 1.$$

$$A^{-1} = \frac{1}{|A|} M' = \begin{pmatrix} -2 & -1 & 3 \\ 1 & 0 & -1 \\ 1 & 2 & -2 \end{pmatrix} .$$

$$\begin{pmatrix} x \\ y \\ z \end{pmatrix} = A^{-1} \begin{pmatrix} 5 \\ 6 \\ 6 \end{pmatrix} = \begin{pmatrix} 2 \\ -1 \\ 5 \end{pmatrix} . \qquad x = 2,\ y = -1,\ z = 5.$$

10. (a) $\frac{1}{60}\begin{pmatrix} -14 & 55 & 80 \\ 10 & -35 & -40 \\ 2 & -25 & -20 \end{pmatrix}$ (b) $\frac{1}{2}\begin{pmatrix} 7 & -2 & 17 \\ -6 & 2 & -16 \\ 7 & -2 & 19 \end{pmatrix}$

(c) $\begin{pmatrix} -25 & 26 & -33 \\ 4 & -4 & 5 \\ 3 & -3 & 4 \end{pmatrix}$.

11. (a) Let S be subset. Let $u, v \in S$.

$$u = (x_1, x_2, x_3, x_1), \qquad v = (x_1', x_2', x_3', x_1').$$

Then $u + v = (x_1 + x_1',\ x_2 + x_2',\ x_3 + x_3',\ x_1 + x_1')$.

Thus $u + v \in S$: Let $a \in \mathbf{R}$.

Then $au = (ax_1, ax_2, ax_3, ax_1) \in S$.

Thus S is a subspace of $\mathbf{R}^4$.

(b) Subspace.

(c) Not subspace. (1,0,0,0) and (0,1,0,0) are in the set but (1,0,0,0) + (0,1,0,0) = (1,1,0,0) is not.

12. $[(f_1 + f_2) + f_3]\ (x) = (f_1 + f_2)(x) + f_3(x) =$

$= (f_1(x) + f_2(x)) + f_3(x) = f_1(x) + (f_2(x) + f_3(x)) =$

$= [f_1 + (f_2 + f_3)]\ (x),\ x \in S$.

Thus $(f_1 + f_2) + f_3 = f_1 + (f_2 + f_3)$.

Associativity for +. $(-f)$ is additive inverse of f, where $(-f)(x) = -(f(x))$.

0 is the zero function, $0(x) = 0$.

Let $a \in \mathbf{R}$. $[a(f_1 + f_2)]\ (x) = a((f_1 + f_2)(x)) =$

$= a(f_1(x) + f_2(x)) = af_1(x) + af_2(x) =$

$= (af_1 + af_2)(x)$, for all $x \in S$.

Thus $a(f_1 + f_2) = af_1 + af_2$. Similarly for other axioms.

(a) Let $U_1 = \{f \in V : f(1) = 0\}$. Let $f_1, f_2 \in U_1$.

Then $(f_1 + f_2)(1) = f_1(1) + f_2(1) = 0 + 0 = 0$.

Thus $f_1 + f_2 \in U_1$. Let $a \in \mathbf{R}$, then $(af)(1) =$

$= af(1) = a\,0 = 0$, if $f \in U_1$. Thus $a\,f \in U_1$.

Hence U_1 is a subspace of V.

(b) Subspace.

(c) Subspace.

(d) Take $f(x) = x^2$, $a = -1$. Then $(af)(x) = -x^2$.

Thus $af \notin$ the subset, but f is in the subset. Hence not a subspace.

(e) Subspace.

13. By inspection, (1,2,0) and (–1,1,0) are linearly independent.

But $(1,1,0) = \frac{2}{3}(1,2,0) + (-\frac{1}{3})(-1,1,0)$.

Thus basis is $\{(1,2,0), (-1,1,0)\}$.

14. Take (1,1,0,–1), then include (1,–1,2,3) and check if (1,0,1,0) is dependent on these two. It is **not**, so include this in the basis. Then check to see if (3,–2,5,7) is dependent on the 3 vectors.

Alternatively, since the given space has dimension at most 3, we know that 3 vectors suffice as a basis.

Thus basis is: $\{(1,1,0,-1), (1,-1,2,3), (1,0,1,0)\}$.

15. (1,1,–1,1), (1,0,1,1), (1,2,1,1) are linearly independent. Since $\mathbf{R}^4$ has dimension 4, we require a vector (x_1, x_2, x_3, x_4) linearly independent of the given three vectors in order to obtain 4 vectors which form a basis of $\mathbf{R}^4$. A few simple trials should turn up a suitable vector. One such is: (1,0,0,0).

Thus a basis for $\mathbf{R}^4$ is:

$\{(1,1,-1,1), (1,0,1,1), (1,2,1,1), (1,0,0,0)\}$.

You could also find (x_1, x_2, x_3, x_4) so that:

$$\begin{vmatrix} x_1 & x_2 & x_3 & x_4 \\ 1 & 1 & -1 & 1 \\ 1 & 0 & 1 & 1 \\ 1 & 2 & 1 & 1 \end{vmatrix} \neq 0.$$ To do this, find a non-vanishing 3 x 3 minor from $\begin{pmatrix} 1 & 1 & -1 & 1 \\ 1 & 0 & 1 & 1 \\ 1 & 2 & 1 & 1 \end{pmatrix}$

and then put the appropriate $x_i = 1$, and all other $x_j = 0, j \neq i$.

16. (a)

$$\begin{pmatrix} 2 & 0 & -1 & 4 \\ 1 & -2 & 0 & 3 \\ 1 & 6 & -2 & -1 \end{pmatrix} \longrightarrow \begin{pmatrix} 1 & -2 & 0 & 3 \\ 2 & 0 & -1 & 4 \\ 1 & 6 & -2 & -1 \end{pmatrix} \longrightarrow$$

$$\longrightarrow \begin{pmatrix} 1 & -2 & 0 & 3 \\ 0 & 4 & -1 & -2 \\ 0 & 8 & -2 & -4 \end{pmatrix} \longrightarrow \begin{pmatrix} 1 & -2 & 0 & 3 \\ 0 & 4 & -1 & -2 \\ 0 & 0 & 0 & 0 \end{pmatrix} \longrightarrow$$

Thus rank = row rank = 2.

(b) Rank = 2.

17. (a)

$$\left.\begin{matrix} 1 & 2 & 3 \\ 2 & 4 & 5 \\ 3 & 5 & 6 \end{matrix}\right| \begin{matrix} 1 & 0 & 0 \\ 0 & 1 & 0 \\ 0 & 0 & 1 \end{matrix} \quad \xrightarrow[\text{Row ③} - 3 \times \text{①}]{\text{Row ②} - 2 \times \text{①}} \quad \left.\begin{matrix} 1 & 2 & 3 \\ 0 & 0 & -1 \\ 0 & -1 & -3 \end{matrix}\right| \begin{matrix} 1 & 0 & 0 \\ -2 & 1 & 0 \\ -3 & 0 & 1 \end{matrix} \longrightarrow$$

$$\xrightarrow{\text{Row ③} - 3 \times \text{②}} \quad \left.\begin{matrix} 1 & 2 & 3 \\ 0 & 0 & -1 \\ 0 & -1 & 0 \end{matrix}\right| \begin{matrix} 1 & 0 & 0 \\ -2 & 1 & 0 \\ 3 & -3 & 1 \end{matrix} \quad \xrightarrow{\text{①} + 3 \times \text{②}} \quad \left.\begin{matrix} 1 & 2 & 0 \\ 0 & 0 & -1 \\ 0 & -1 & 0 \end{matrix}\right| \begin{matrix} -5 & 3 & 0 \\ -2 & 1 & 0 \\ 3 & -3 & 1 \end{matrix} \longrightarrow$$

$$\text{①} + 2 \times \text{③} \quad \left.\begin{matrix} 1 & 0 & 0 \\ 0 & 0 & -1 \\ 0 & -1 & 0 \end{matrix}\right| \begin{matrix} 1 & -3 & 2 \\ -2 & 1 & 0 \\ 3 & -3 & 1 \end{matrix} \quad \xrightarrow[\text{② and ③ Change signs}]{\text{Interchange}} \quad \left.\begin{matrix} 1 & 0 & 0 \\ 0 & 1 & 0 \\ 0 & 0 & 1 \end{matrix}\right| \begin{matrix} 1 & -3 & 2 \\ -3 & 3 & -1 \\ 2 & -1 & 0 \end{matrix}.$$

Thus

$$\begin{pmatrix} 1 & 2 & 3 \\ 2 & 4 & 5 \\ 3 & 5 & 6 \end{pmatrix}^{-1} = \begin{pmatrix} 1 & -3 & 2 \\ -3 & 3 & -1 \\ 2 & -1 & 0 \end{pmatrix}.$$

(b) $\begin{pmatrix} 7 & -3 & -3 \\ -1 & 1 & 0 \\ -1 & 0 & 1 \end{pmatrix}$ (c) $\begin{pmatrix} 1 & 0 & 0 \\ 2 & -1 & 2 \\ 1 & 0 & 1 \end{pmatrix}$.

18. (a)
$$\left.\begin{matrix} 2 & 4 & 1 \\ 3 & 5 & 0 \\ 5 & 13 & 7 \end{matrix}\right| \begin{matrix} 1 \\ 1 \\ 4 \end{matrix} \quad \xrightarrow[③ - 2 \times ①]{② - ①} \quad \left.\begin{matrix} 2 & 4 & 1 \\ 1 & 1 & -1 \\ 1 & 5 & 5 \end{matrix}\right| \begin{matrix} 1 \\ 0 \\ 2 \end{matrix} \quad \longrightarrow$$

$$\xrightarrow{① - 2 \times ②} \quad \left.\begin{matrix} 0 & 2 & 3 \\ 1 & 1 & -1 \\ 0 & 4 & 6 \end{matrix}\right| \begin{matrix} 1 \\ 0 \\ 2 \end{matrix} \quad \xrightarrow{③ - 2 \times ①} \quad \left.\begin{matrix} 0 & 2 & 3 \\ 1 & 1 & -1 \\ 0 & 0 & 0 \end{matrix}\right| \begin{matrix} 1 \\ 0 \\ 0 \end{matrix}.$$

Take $x_3 = \lambda$ Then $x_2 = \frac{1}{2}(1 - 3\lambda)$, $x_1 = -\frac{1}{2} + \frac{5\lambda}{2}$.

Thus
$$\begin{pmatrix} x_1 \\ x_2 \\ x_3 \end{pmatrix} = \begin{pmatrix} -\frac{1}{2} \\ \frac{1}{2} \\ 0 \end{pmatrix} + \lambda \begin{pmatrix} \frac{5}{2} \\ \frac{-3}{2} \\ 1 \end{pmatrix}.$$

$\left\langle \begin{pmatrix} \frac{5}{2} \\ \frac{-3}{2} \\ 1 \end{pmatrix} \right\rangle$ is the solution space of the homogeneous system $A\boldsymbol{x} = \boldsymbol{0}$.

$\begin{pmatrix} -\frac{1}{2} \\ \frac{1}{2} \\ 0 \end{pmatrix}$ is a particular solution of the non-homogeneous system as given.

Compare the form of the solution with that for linear differential equations.

(b) Solution space $\langle(-14, 11, 16)'\rangle$.

(c) No solution.

(d) $$\begin{pmatrix} 2 \\ -1 \\ 1 \\ 0 \end{pmatrix} + \lambda \begin{pmatrix} 12 \\ 1 \\ -20 \\ 7 \end{pmatrix}, \quad \lambda \text{ arbitrary.}$$

(e) $$\begin{pmatrix} 0 \\ 1 \\ -1 \\ 1 \end{pmatrix} + \lambda \begin{pmatrix} 1 \\ -3 \\ 0 \\ 0 \end{pmatrix}, \quad \lambda \text{ arbitrary.}$$

19. (a) Take $v, v_1, v_2 \in \mathbf{R}^3$, and $a \in \mathbf{R}$.

Then show that $f(v_1 + v_2) = f(v_1) + f(v_2)$ and $f(av) = a f(v)$.

Tedious but straight forward.

(b) $f((x_1, x_2, x_3)) = (x_1 + x_2 + x_3,\ x_1 - x_2 - x_3) = (0,0)$.

Solve $x_1 + x_2 + x_3 = 0$. $\quad \operatorname{Ker} f = \langle (0,1,-1) \rangle$

$x_1 - x_2 - x_3 = 0$. $\quad \dim \operatorname{Ker} f = 1$.

(c) $\dim \mathbf{R}^3 = \dim \operatorname{Ker} f + \dim \operatorname{Im} f$. Thus $3 = 1 + \dim \operatorname{Im} f$.

Hence $\dim \operatorname{Im} f = 2$. $\quad$ But $\operatorname{Im} f \subset \mathbf{R}^2$ and $\dim \mathbf{R}^2 = 2$.

Hence $\operatorname{Im} f = \mathbf{R}^2$.

(d) $$A = \frac{1}{5} \begin{pmatrix} 4 & -2 & 9 \\ 2 & 4 & -3 \end{pmatrix}.$$

20. (a) As for 19 (a).

(b) $v \in \operatorname{Ker} f \leftrightarrow f(v) = 0$. Let $v = (x, y, z)$.

Then $f((x, y, z)) = (x + y,\ y - z,\ x + z) = (0,0,0)$.

Thus $\operatorname{Ker} f = \langle (-1,1,1) \rangle$. $\quad \dim \operatorname{Ker} f = 1$.

(c) $\operatorname{Im} f = \langle f((1,0,0)),\ f((0,1,0)),\ f((0,0,1)) \rangle$,

because $(1,0,0), (0,1,0), (0,0,1)$ generate $\mathbf{R}^3$.

Thus $\quad \operatorname{Im} f = \langle (1,0,1), (1,1,0), (0,-1,1) \rangle$.

Now $\quad (0,-1,1) = (1,0,1) + (-1)(1,1,0)$.

Thus $\quad \operatorname{Im} f = \langle (1,0,1), (1,1,0) \rangle$.

By inspection (1,0,1), (1,1,0) are linearly independent, so form a basis of $\operatorname{Im} f$.

Thus $\quad \dim \operatorname{Im} f = 2$.

Hence $\quad \dim \operatorname{Ker} f + \dim \operatorname{Im} f = 1 + 2 = 3 = \dim \mathbf{R}^3$.

(d) $f((1,0,0)) = (1,0,1) = 2(1,0,0) + 0(0,-1,2) + 1(-1,0,1)$

$f((0,-1,2)) = (-1,-3,2) = -5(1,0,0) + 3(0,-1,2) + (-4)(-1,0,1)$

$f((-1,0,1)) = (-1,-1,0) = -3(1,0,0) + 1(0,-1,2) + (-2)(-1,0,1)$

Matrix representing f is $A = \begin{pmatrix} 2 & -5 & -3 \\ 0 & 3 & 1 \\ 1 & -4 & -2 \end{pmatrix}$.

(e) Similarly $B = \begin{pmatrix} \frac{5}{4} & -\frac{3}{4} & -\frac{1}{2} \\ \frac{3}{4} & \frac{3}{4} & \frac{3}{2} \\ \frac{1}{2} & \frac{1}{2} & 1 \end{pmatrix}$.

(f) To find P write basis $\{(1,0,0), (0,-1,2), (-1,0,1)\}$ in terms of basis $\{(1,1,0), (1,-1,0), (0,1,2)\}$.

Thus $\quad (1,0,0) = \frac{1}{2}(1,1,0) + \frac{1}{2}(1,-1,0) + 0(0,1,2)$

$(0,-1,2) = -1(1,1,0) + 1(1,-1,0) + 1(0,1,2)$

$(-1,0,1) = \frac{-3}{4}(1,1,0) + (\frac{-1}{4})(1,-1,0) + \frac{1}{2}(0,1,2)$.

Hence $\quad P = \begin{pmatrix} \frac{1}{2} & -1 & -\frac{3}{4} \\ \frac{1}{2} & 1 & -\frac{1}{4} \\ 0 & 1 & \frac{1}{2} \end{pmatrix}$.

By calculation we get

$$P^{-1} = \begin{pmatrix} 3 & -1 & 4 \\ -1 & 1 & -1 \\ 2 & -2 & 4 \end{pmatrix}.$$

A direct calculation verifies that $P^{-1}BP = A$.

(g) $\text{Im} f$ has dimension 2, hence rank $A = 2 =$ rank B. This can also be verified by direct calculation. P has an inverse so $|P| \neq 0$. Thus rank $P = 3$.

21. Take basis $\{(1,0,0), (0,1,0), (0,0,1)\}$ of $\mathbf{R}^3$.

Define $f : \mathbf{R}^3 \to \mathbf{R}^4$ by:

$$f((1,0,0)) = (1,2,0,-4)$$

$$f((0,1,0)) = (2,0,-1,-3)$$

$$f((0,0,1)) = (0,0,0,0),$$

and
$$\begin{aligned} f((x,y,z)) &= f(x(1,0,0) + y(0,1,0) + z(0,0,1)) \\ &= xf((1,0,0)) + yf((0,1,0)) + zf((0,0,1)) \\ &= x(1,2,0,-4) + y(2,0,-1,-3) \\ &= (x + 2y,\ 2x,\ -y,\ -4x - 3y). \end{aligned}$$

22. (a) Let $A, B \in V$. Then $f(A + B) = (A + B)M - M(A + B) =$

$$= AM + BM - MA - MB = (AM - MA) + (BM - MB) =$$

$$= f(A) + f(B).$$ Let $a \in \mathbf{R}$.

Then $f(aA) = (aA)M - M(aA) = a(AM - MA) = af(A)$.

Thus f is a linear transformation on V.

(b) $A \in \text{Ker} f \leftrightarrow f(A) = \begin{pmatrix} 0 & 0 \\ 0 & 0 \end{pmatrix} \leftrightarrow$

$$\leftrightarrow \begin{pmatrix} a & b \\ c & d \end{pmatrix} \begin{pmatrix} 1 & 2 \\ 0 & 3 \end{pmatrix} - \begin{pmatrix} 1 & 2 \\ 0 & 3 \end{pmatrix} \begin{pmatrix} a & b \\ c & d \end{pmatrix} = \begin{pmatrix} 0 & 0 \\ 0 & 0 \end{pmatrix} \leftrightarrow$$

$$\leftrightarrow \begin{pmatrix} -2c & 2a+2b-2d \\ -2c & 2c \end{pmatrix} = \begin{pmatrix} 0 & 0 \\ 0 & 0 \end{pmatrix} \leftrightarrow$$

$\leftrightarrow \; c = 0, \quad a + b - d = 0.$

Take $b = \lambda \;\; d = \mu$, where λ, μ are arbitrary in $\mathbf{R}$.

Then $c = 0, \quad a = -\lambda + \mu$.

Thus $A \in \text{Ker } f \leftrightarrow A = \begin{pmatrix} -\lambda + \mu & \lambda \\ 0 & \mu \end{pmatrix} =$

$$= \mu \begin{pmatrix} 1 & 0 \\ 0 & 1 \end{pmatrix} + \lambda \begin{pmatrix} -1 & 1 \\ 0 & 0 \end{pmatrix}.$$

Thus $\text{Ker } f = \left\langle \begin{pmatrix} 1 & 0 \\ 0 & 1 \end{pmatrix}, \begin{pmatrix} -1 & 1 \\ 0 & 0 \end{pmatrix} \right\rangle$; $\dim \text{Ker } f = 2$.

(c) A basis of V is: $\left\{ \begin{pmatrix} 1 & 0 \\ 0 & 0 \end{pmatrix}, \begin{pmatrix} 0 & 1 \\ 0 & 0 \end{pmatrix}, \begin{pmatrix} 0 & 0 \\ 1 & 0 \end{pmatrix}, \begin{pmatrix} 0 & 0 \\ 0 & 1 \end{pmatrix} \right\}$

Thus $\dim V = 4$. Now $\dim V = \dim \text{Ker } f + \dim \text{Im } f$.

Hence $\text{Im } f$ has dimension 2.

23. $(f + g)((x,y,z)) = f((x,y,z)) + g((x,y,z)) =$

$$= (2x, y + z) + (x - z, y)$$

$$= (3x - z, 2y + z).$$

Similarly $(3f)((x,y,z)) = (6x, 3y + 3z)$, and

$$(2f - 5g)((x,y,z)) = (-x + 5z, -3y + 2z).$$

24. Suppose $af + bg + ch = 0$ for some $a,b,c \in \mathbf{R}$.

Take basis $\{(1,0,0), (0,1,0), (0,0,1)\}$ of $\mathbf{R}^3$.

Then $\quad (af + bg + ch)((1,0,0)) = (0,0)$.

Thus $\quad (a + 2b, a+b+c) = (0,0)$.

Hence $a + 2b = 0$, $a+b+c = 0$. Repeat for (0,1,0).

Then $a + 2c = 0$, $a+b = 0$. Altogether $a = b = c = 0$.

Required result follows.

Exercises 5

1. (a) Direction is given by vector: $(4,2,3) - (1,1,1) = (3,1,2)$.
Equation in vector form is: $\boldsymbol{r} = (1,1,1) + \lambda(3,1,2)$.
Equation in symmetric form is: $(x-1)/_3 = (y-1)/_1 = (z-1)/_2$,
which can be written down at once from the vector form and conversely.

(b) $\boldsymbol{r} = (2,0,1) + \lambda(5,4,1)$; $(x-2)/_5 = y/_4 = (z-1)/_1$.

(c) $\boldsymbol{r} = (1,4,2) + \lambda(-1,1,1)$; $(x-1)/_{-1} = (y-4)/_1 = (z-2)/_1$.

2. Angle between lines (a) and (b): let $3\boldsymbol{i} + \boldsymbol{j} + 2\boldsymbol{k}$ be a vector in the direction of line (a) with the usual unit vectors $\boldsymbol{i}$, $\boldsymbol{j}$, $\boldsymbol{k}$. Let $5\boldsymbol{i} + 4\boldsymbol{j} + \boldsymbol{k}$ be a vector in direction of line (b). Then

$$(3\boldsymbol{i} + \boldsymbol{j} + 2\boldsymbol{k}) \cdot (5\boldsymbol{i} + 4\boldsymbol{j} + \boldsymbol{k}) = |3\boldsymbol{i} + \boldsymbol{j} + 2\boldsymbol{k}||5\boldsymbol{i} + 4\boldsymbol{j} + \boldsymbol{k}| \cos\theta.$$

where θ is the angle between the lines. Thus

$$\cos\theta = (15 + 4 + 2)/\sqrt{14}\cdot\sqrt{42} = 21/14\sqrt{3} = \sqrt{3}/2.$$

We take $\theta = 30$ degrees. (Unless the lines are directed we take acute angles).

Angle between lines (a) and (c) is: $\cos^{-1} 0 = 90$ degrees.

Angles between lines (b) and (c) is: $\cos^{-1} 0 = 90$ degrees.

3. (a) $\boldsymbol{r} = (0,1,-1) + \lambda(1,2,3)$; $x/_1 = (y-1)/_2 = (z+1)/_3$.

(b) $\boldsymbol{r} = (-1,0,0) + \lambda(1,-1,1)$; $(x+1)/_1 = y/_{-1} = z/_1$.

(c) $\boldsymbol{r} = (1,1,1) + \lambda(2,-1,1)$; $(x-1)/_2 = (y-1)/_{-1} = (z-1)/_1$.

4. (a) Let P be point $(1, 4+\lambda, \lambda)$ and

Q be $(1 + 2\mu, -1 + \mu, 4 + 2\mu)$. Segment $\overrightarrow{PQ}$ is: $(2\mu, \mu - \lambda - 5, 4 + 2\mu - \lambda)$.

$\overrightarrow{PQ}$ must be perpendicular to both lines.

Thus $\overrightarrow{PQ} \cdot (0,1,1) = 0, \quad \overrightarrow{PQ} \cdot (2,1,2) = 0.$

Thus $3\mu - 2\lambda = 1, \quad 9\mu - 3\lambda = -3.$

Hence $\lambda = -2, \quad \mu = -1.$

Thus the common perpendicular meets the lines at the points $P = (1,2,-2)$ and $Q = (-1,-2,2)$. The shortest distance between the lines is

$$|\overrightarrow{PQ}| = \sqrt{4 + 16 + 16} = \sqrt{36} = 6.$$

Equation of PQ is: $(x - 1)/_2 = (y - 2)/_4 = (z + 2)/_{-4}$.

(b) Point P on $x/_1 = (y - 1)/_1 = (z - 1)/_1$ can be written $(\lambda, 1+\lambda, 1+\lambda)$.

Point Q on $(x - 1)/_1 = y/_2 = z/_{-3}$ can be written $(1+\mu, 2\mu, -3\mu)$.

Then proceed as above.

We have: $P = (\frac{-1}{3}, \frac{2}{3}, \frac{2}{3}),$

$Q = (\frac{6}{7}, \frac{-2}{7}, \frac{3}{7}).$

Shortest distance $|\overrightarrow{PQ}| = 5\sqrt{42}/_{21}$.

Equation of PQ is: $(x + \frac{1}{3})/_5 = (y - \frac{2}{3})/_{-4} = (z - \frac{2}{3})/_{-1}$.

5. $l^2 + m^2 + n^2 = (l + \delta l)^2 + (m + \delta m)^2 + (n + \delta n)^2 = 1,$

since the sum of squares of direction cosines is 1.

Thus $2(l\,\delta l + m\,\delta m + n\,\delta n) + (\delta l)^2 + (\delta m)^2 + (\delta n)^2 = 0 \leftarrow *$.

Also $\cos \delta\theta = l(l + \delta l) + m(m + \delta m) + n(n + \delta n)$

and $\cos \delta\theta = 1 - (\delta\theta)^2/_2$, approximately.

Thus $1 - (\delta\theta)^2/_2 = (l^2 + m^2 + n^2) + (l\,\delta l + m\,\delta m + n\,\delta n) =$

$$= 1 - \tfrac{1}{2}((\delta l)^2 + (\delta m)^2 + (\delta n)^2), \text{ by } *.$$

Hence $(\delta\theta)^2 = (\delta l)^2 + (\delta m)^2 + (\delta n)^2$.

Thus $\delta\theta = \pm\sqrt{(\delta l)^2 + (\delta m)^2 + (\delta n)^2}$

6. Take $v = l\boldsymbol{i} + m\boldsymbol{j} + n\boldsymbol{k}$ and $u = l'\boldsymbol{i} + m'\boldsymbol{j} + n'\boldsymbol{k}$. The vector product $u \wedge v$ gives a vector in the direction perpendicular to L and M.

Now $u \wedge v = (m'n - n'm)\,\boldsymbol{i} + (n'l - l'n)\boldsymbol{j} + (l'm - m'l)\,\boldsymbol{k}$.

Take N to be a line with direction ratios:

$$(m'n - n'm) : (n'l - l'n) : (l'm - m'l).$$

7. Line joining (5,–3,1) and (3,2,7) is:

$$\boldsymbol{r} = (3,2,7) + \lambda(-2,5,6).$$

This cuts the xy coordinate plane when $z = 0$. That is when $7 + 6\lambda = 0$. Thus $\lambda = \frac{-7}{6}$. The coordinates of the point where line cuts xy plane are:

$(\frac{16}{3}, \frac{-23}{6}, 0)$. $y = 0$ when $2 + 5\lambda = 0$, $\lambda = \frac{-2}{5}$.

Thus coordinates of the point where the line cuts xz plane are:

$(\frac{19}{5}, 0, \frac{23}{5})$. $x = 0$ when $3 - 2\lambda = 0$, $\lambda = 3/2$.

Coordinates of the point where line cuts yz plane are:

$(0, \frac{19}{2}, 16)$. We want the ratio in which line is divided by point

$(\frac{19}{5}, 0, \frac{23}{5})$.

We need look only at the y coordinates of the three points:

$(5,-3,1)$, $(\frac{19}{5}, 0, \frac{23}{5})$, $(3,2,7)$

2
0
−3

Thus the required ratio is 3 : 2.

8. If we expand the given determinant we get a linear equation in x, y, z, i.e. the equation of a plane. If we put x_i, y_i, z_i in the place of x, y, z respectively in the first row, then two rows of the determinant become equal so that the determinant vanishes for $i = 2, 3$. For $i = 1$, the first row is zero and the determinant vanishes. This means that the three given points lie on the plane given by the determinantal equation, as required.

9. Normals to planes are $u = 3\boldsymbol{i} - 2\boldsymbol{j} + \boldsymbol{k}$ and $v = 5\boldsymbol{i} + 4\boldsymbol{j} - 6\boldsymbol{k}$. Line of intersection has direction $u \wedge v = 8\boldsymbol{i} + 23\boldsymbol{j} + 22\boldsymbol{k}$.

Solve $3x - 2y + z = 1$

$5x + 4y - 6z = 2$ for a common point,

say $(\frac{4}{11}, \frac{1}{22}, 0)$, or $(0,-1,-1)$.

Then line of intersection is:

$$\boldsymbol{r} = (0,-1,-1) + \lambda(8,23,22), \text{ or}$$

$$\frac{x}{8} = (y+1)/_{23} = (z+1)/_{22}.$$

10. Distance from (0,1,1) to the plane $x + y + z = 1$. Line through (0,1,1) perpendicular to the plane is:

$\boldsymbol{r} = (0,1,1) + \lambda(1,1,1)$. This cuts plane where λ is given by:

$\lambda + (1 + \lambda) + (1 + \lambda) = 1$. Thus $\lambda = -\frac{1}{3}$.

Point is: $(-\frac{1}{3}, \frac{2}{3}, \frac{2}{3})$. Required distance is between (0,1,1) and $(-\frac{1}{3}, \frac{2}{3}, \frac{2}{3})$.

This is: $\sqrt{(\frac{1}{3})^2 + (\frac{1}{3})^2 + (\frac{1}{3})^2} = \frac{\sqrt{3}}{3}$.

11. Using the result of exercise 8, the required planes are as follows:

(a) $$\begin{vmatrix} x-1 & y-2 & z-3 \\ 3-1 & 5-2 & 7-3 \\ 3-1 & -1-2 & -3-3 \end{vmatrix} = 0 \longrightarrow \begin{vmatrix} x-1 & y-2 & z-3 \\ 2 & 3 & 4 \\ 2 & -3 & -6 \end{vmatrix} = 0$$

Plane is: $3x - 10y + 6z = 1$.

(b) $7x + 13y + 14z = 75$.

(c) $53x + 5y - 3z = 54$.

12. Use the method of exercise 9 and note that (1,2,3) lies on all three planes. Then we have:

Line of intersection of (a) and (b) is:

$$(x-1)/_{-2} = (y-2)/_{0} = (z-3)/_{1}.$$

Line of intersection of (a) and (c) is:

$$(x - 1)/0 = (y - 2)/3 = (z - 3)/5.$$

Line of intersection of (b) and (c) is:

$$(x - 1)/1 = (y - 2)/{-7} = (z - 3)/6.$$

The planes intersect in the one point (1,2,3).

13. Take the given point $P = (-2,1,3)$ and two points, say (1,2,5) and (–2,–2,–1), which lie on the line L. Then use exercise 8. Required plane is:

$$\begin{vmatrix} x+2 & y-1 & z-3 \\ 1+2 & 2-1 & 5-3 \\ -2+2 & -2-1 & -1-3 \end{vmatrix} = 0. \text{ This gives: } 2x + 12y - 9z + 19 = 0.$$

The plane through L perpendicular to the above plane has its normal perpendicular to L and to the normal to the plane $2x + 12y - 9z + 19 = 0$ just found.

By exercise 6, or using a suitable vector product, the direction ratios of the normal to the required plane are : $-108 : 39 : 28$. The point (1,2,5) lies on L and hence must lie on the plane. Thus the required plane is:

$$-108(x - 1) + 39(y - 2) + 28(z - 5) = 0.$$

This simplifies to: $108x - 39y - 28z + 110 = 0.$

14. By exercise 6, the direction of the line of intersection of $x + y - z = 1$ and $2x - 3y + z = 2$ is : $-2 : -3 : -5$ or $2 : 3 : 5$. The direction ratios of the line joining (3,–1,2) and (4,0,–1) are $1 : 1 : -3$. Let $u = 2\boldsymbol{i} + 3\boldsymbol{j} + 5\boldsymbol{k}$ and $v = \boldsymbol{i} + \boldsymbol{j} - 3\boldsymbol{k}$ be vectors in these directions. Let θ be the angle between these lines.

Then $u \cdot v = |u||v| \cos \theta.$

Thus $\theta = \cos^{-1} u \cdot v/|u||v| = \cos^{-1}(2+3-15)/\sqrt{38}\sqrt{11} =$

$$= \cos^{-1} -10/\sqrt{418}$$

The acute angle is then $\cos^{-1} {}^{10}/\sqrt{418} = 60.72$ degrees approximately.

$u \wedge v$ is a vector perpendicular to both lines and $u \wedge v = -14i + 11j - k$. Thus the direction ratios of the line perpendicular to both lines is : $-14 : 11 : -1$. Required line through origin in this direction is : $r = (0,0,0) + \lambda(-14,11,-1)$, or

$$x/{-14} = y/11 = z/{-1}.$$

15. The new plane will contain the line of intersection L of the 2 given planes. Moreover the normal to the new plane will be perpendicular to L and also to the normal to the plane $4x + 7y + 4z + 81 = 0$.

First find L. Use exercise 6 to find direction ratios of the line L. They are $58 : -20 : -23$. A point on both planes is: $(-\frac{91}{2}, 0, \frac{101}{4})$.

Thus L is: $r = (-\frac{91}{2}, 0, \frac{101}{4}) + \lambda(58,-20,-23)$.

The normal to the new plane will be perpendicular to L and to the direction $4 : 7 : 4$. Using exercise 6 again, the normal to the new plane has direction ratios $1 : -4 : 6$. A point on L is $(-91/2, 0, 101/4)$.

New plane has equation $1 \cdot (x + \frac{91}{2}) - 4(y - 0) + 6(z - \frac{101}{4}) = 0$.

This reduces to: $x - 4y + 6z = 106$.

Line through origin perpendicular to plane

$$4x + 7y + 4z + 81 = 0 \text{ is: } r = (0,0,0) + \lambda(4,7,4).$$

This meets plane when $4(4\lambda) + 7(7\lambda) + 4(4\lambda) = -81$, thus $\lambda = -1$.

Foot of this perpendicular is: $(-4,-7,-4)$.

Line through origin perpendicular to plane

$$x - 4y + 6z = 106 \text{ is: } r = (0,0,0) + \mu(1,-4,6).$$

This meets plane when $\mu - 4(-4\mu) + 6(6\mu) = 106$. Thus $\mu = 2$.

Foot of perpendicular is: $(2,-8,12)$.

Distance between feet of perpendiculars is:

$$\sqrt{6^2 + (-1)^2 + (16)^2} = \sqrt{293}.$$

16. Sphere is: $(x-1)^2+(y+2)^2+(z-3)^2 = 4$ or

$$x^2+y^2+z^2-2x+4y-6z+10 = 0.$$

17. $2(x^2+y^2+z^2)-6x+8y-8z = 1$ can be written

$$x^2+y^2+z^2-3x+4y-4z = \tfrac{1}{2} \quad \rightarrow$$

$$\rightarrow (x-\tfrac{3}{2})^2 + (y+2)^2 + (z-2)^2 = \tfrac{9}{4}+4+4+\tfrac{1}{2} \quad \rightarrow$$

$$(x-\tfrac{3}{2})^2 + (y+2)^2 + (z-2)^2 = (\tfrac{\sqrt{43}}{2})^2.$$

Thus centre is: $(\tfrac{3}{2}, -2, 2)$ and radius is: $\sqrt{43}/2$.

18. Let equation of sphere be: $(x^2+y^2+z^2)+ax+by+cz+d = 0$.
If sphere has to go through the points (x_1, y_1, z_1), (x_2, y_2, z_2), (x_3, y_3, z_3), (x_4, y_4, z_4), we have 5 equations as follows:

$$1 \cdot (x^2+y^2+z^2)+ax+by+cz+d = 0.$$

$$1 \cdot (x_i^2+y_i^2+z_i^2)+ax_i+by_i+cz_i+d = 0; \quad i = 1,2,3,4.$$

If we regard these equations as 5 linear homogeneous equations in the 'variables' 1, a, b, c, d, which are **not** all zero, the necessary and sufficient condition for the existence of such a system is that the determinant of the coefficients of 1, a, b, c, d, should vanish. This gives:

$$\begin{vmatrix} x^2+y^2+z^2 & x & y & z & 1 \\ x_1^2+y_1^2+z_1^2 & x_1 & y_1 & z_1 & 1 \\ x_2^2+y_2^2+z_2^2 & x_2 & y_2 & z_2 & 1 \\ x_3^2+y_3^2+z_3^2 & x_3 & y_3 & z_3 & 1 \\ x_4^2+y_4^2+z_4^2 & x_4 & y_4 & z_4 & 1 \end{vmatrix}$$

This is the equation of a sphere, provided the 4 given points are not coplanar.

Using this method in our case gives:

(a)
$$\begin{vmatrix} x^2+y^2+z^2 & x & y & z & 1 \\ 3 & 1 & -1 & 1 & 1 \\ 5 & 0 & 1 & 2 & 1 \\ 13 & 2 & 3 & 0 & 1 \\ 45 & 5 & 2 & 4 & 1 \end{vmatrix}$$

This leads to: $x^2 + y^2 + z^2 - 6x - 2y - 4z + 5 = 0$.

Note that the equation above in determinantal form is a sphere type equation, which is satisfied by the 4 given points, since, on substitution of each point for (x,y,z) in the first row, two rows become equal, so the determinant vanishes. Thus the determinantal equation does give the required sphere when it exists, i.e. always when the 4 points are **not** coplanar, and whenever 4 coplanar points do have a sphere through them.

(b) In a similar way the required sphere is:

$$x^2 + y^2 + z^2 - 2x - 2y - 2z - 14 = 0.$$

An alternative approach, more geometrical, is to find the 3 planes which are perpendicular bisectors of the lines connecting the first point (say) to each of the other 3 points. The centre is the point of intersection of these 3 planes. The radius is the distance from the centre to one of the given points. With the centre and radius found, it is simple to find the equation of the required sphere.

19. The direction ratios of the radius of the sphere at $(1, -2, 2)$ are $1 : -2 : 2$, because radius connects $(0, 0, 0)$ to $(1, -2, 2)$. Thus the tangent plane is a plane whose normal has direction ratios $1 : -2 : 2$ and contains the point $(1, -2, 2)$.

Tangent plane is: $1 \cdot (x - 1) + (-2)(y + 2) + 2(z - 2) = 0$.

This reduces to : $x - 2y + 2z = 9$.

20. Sphere is: $(x + 1)^2 + (y - 2)^2 + (z + 3)^2 = (\sqrt{21})^2$. Centre at $(-1, 2, -3)$. Any plane through the given line is: $(6x - 3y - 23) + \lambda\,(3z + 2) = 0$ or $3z + 2 = 0$, where λ is arbitrary. If one of these planes is to be a tangent plane, its distance from the centre of the sphere must equal the radius $\sqrt{21}$. The distance is:

$$\pm\ [(-6 - 6 - 23) + \lambda(-9 + 2)] / \sqrt{36 + 9 + 9\lambda^2} =$$

$$= \pm\ (7\lambda + 35) / 3\sqrt{5 + \lambda^2}.$$

Thus $\pm\ (7\lambda + 35) = 3\sqrt{21}\sqrt{5 + \lambda^2}$. Hence $2\lambda^2 - 7\lambda - 4 = 0$.

$$(2\lambda + 1)(\lambda - 4) = 0;\ \lambda = 4, \text{ or } \tfrac{-1}{2}.$$

The distance from $(-1, 2, -3)$ to the plane $3z + 2 = 0$ is:

$$\left| (-9+2)/\sqrt{9} \right| = \tfrac{7}{3} \neq \sqrt{21}.$$

Hence there are 2 tangent planes intersecting the given line; namely

$$(6x - 3y - 23) + 4(3z + 2) = 0 \text{ and } (6x - 3y - 23) - \tfrac{1}{2}(3z + 2) = 0.$$

These reduce to: $4x - 2y - z = 16$ and $2x - y + 4z = 5$.

21. The given circle is that in which the sphere $x^2 + y^2 + z^2 - 2x + 2vy + 2z - 2 = 0$ is cut by the plane $y = 0$. This sphere touches the plane $-y + z + 7 = 0$ if the perpendicular from $(1, -v, -1)$ to the plane is equal to the radius. Here $(1, -v, -1)$ is the centre of the above sphere which may be written:

$$(x-1)^2 + (y+v)^2 + (z+1)^2 = 2 + 1^2 + v^2 + 1^2 = 4 + v^2 =$$

$$= (\sqrt{4+v^2})^2.$$

The radius is $\sqrt{4+v^2}$. Hence the sphere touches the given plane if

$$\pm(v - 1 + 7)/\sqrt{1^2 + 1^2} = \sqrt{4+v^2}.$$

This leads to: $v^2 - 12v - 28 = 0$. Hence $v = 14$ or -2.

Thus required spheres are: $x^2 + y^2 + z^2 - 2x + 28y + 2z - 2 = 0$

and $x^2 + y^2 + z^2 - 2x - 4y + 2z - 2 = 0$.

22. Because of the symmetry of this particular transformation, it is possible to see 'at a glance' the required results. However we give below a much longer and more tedious approach which is suitable in general.

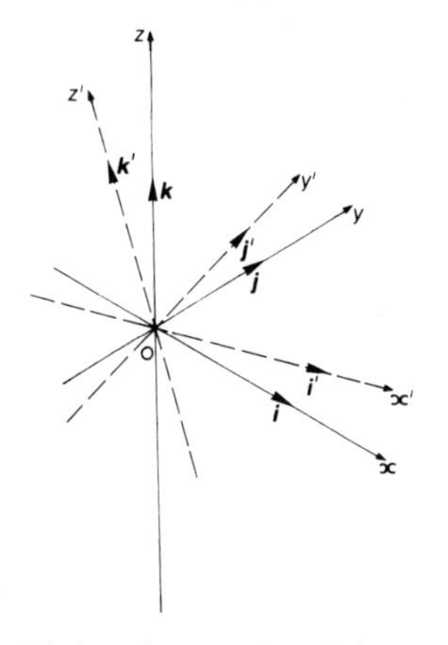

The rotation takes the $Oxyz$ axes into the $Ox'y'z'$ axes. If f denotes the transformation involved, we have:

$$f(i) = i', f(j) = j', f(k) = k'.$$

If A is the matrix of the coordinate transformation we have:

$$\begin{pmatrix} x' \\ y' \\ z' \end{pmatrix} = A \begin{pmatrix} x \\ y \\ z \end{pmatrix} \quad \text{and} \quad \begin{aligned} i' &= a_{11}\, i + a_{12}\, j + a_{13}\, k \\ j' &= a_{21}\, i + a_{22}\, j + a_{23}\, k \\ k' &= a_{31}\, i + a_{32}\, j + a_{33}\, k \end{aligned}$$

Note that f preserves lengths so $|f(u)| = |u|$.

Consider the plane formed by the axis of rotation in the direction $\frac{1}{\sqrt{3}}(i+j+k)$ and the vector i.

In this plane we write i as the sum of a vector v along the axis of rotation and a vector u perpendicular to the axis of rotation. We find that $v = \frac{1}{3}(i+j+k)$ and $u = \frac{1}{3}(2i - j - k)$. Now $f(v) = v$ because v is along the axis of rotation.

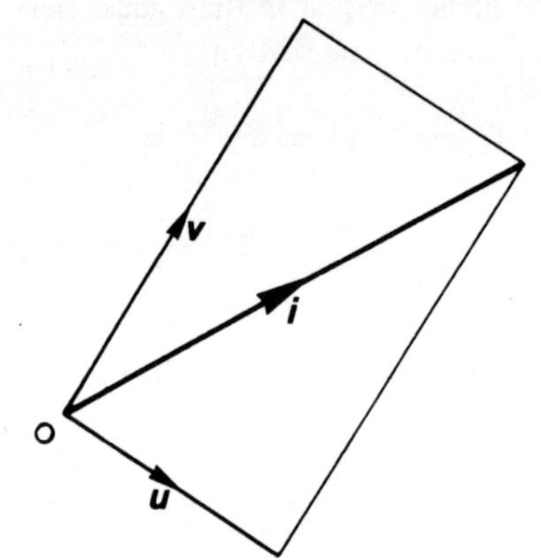

Also $f(u) \cdot u = |f(u)|\,|u| \cos 120° =$

$= |u|^2 (-\frac{1}{2}) = \frac{6}{9}(-\frac{1}{2})$.

Thus $\quad f(u) \cdot u = -\frac{1}{3}, \quad f(u) \cdot v = 0, \quad \text{and} \quad f(u) \cdot f(u) = \frac{6}{9}$.

Let $\quad f(u) = ai + bj + ck$. Then $a^2 + b^2 + c^2 = \frac{2}{3}, \quad 2a - b - c = -1,$

$a + b + c = 0.$

Hence $\quad f(u) = \frac{1}{3}(-i + 2j - k)$ **or** $f(u) = \frac{1}{3}(-i - j + 2k)$.

Counter-clockwise rotation gives positive coefficient of k.

Hence we take $\quad f(u) = \frac{1}{3}(-i - j + 2k)$.

Now $\quad f(i) = f(u+v) = f(u) + f(v) =$

$$= \frac{1}{3}(-i - j + 2k) + \frac{1}{3}(i + j + k).$$

Thus $\quad f(i) = k$. Hence $\quad a_{11}\, i + a_{12}\, j + a_{13}\, k = k$.

Similarly $\quad f(j) = i$ and $f(k) = j$.

Thus $\quad a_{21}\, i + a_{22}\, j + a_{23}\, k = i, \quad a_{31}\, i + a_{32}\, j + a_{33}\, k = j.$

Thus the matrix A is: $\begin{pmatrix} 0 & 0 & 1 \\ 1 & 0 & 0 \\ 0 & 1 & 0 \end{pmatrix}$

The matrix is orthogonal because $A'A = I_3$.

$$|A| = 1 . A^3 = \begin{pmatrix} 0 & 0 & 1 \\ 1 & 0 & 0 \\ 0 & 1 & 0 \end{pmatrix} \begin{pmatrix} 0 & 1 & 0 \\ 0 & 0 & 1 \\ 1 & 0 & 0 \end{pmatrix} = \begin{pmatrix} 1 & 0 & 0 \\ 0 & 1 & 0 \\ 0 & 0 & 1 \end{pmatrix} =$$

$$= I_3, \text{ as required.}$$

23. The given matrix A is orthogonal and $|A| = 1$. Thus A preserves lengths and distances. Moreover

$$|A - I_3| = |A' - I_3| = |A^{-1} - I_3| = |A^{-1}(I_3 - A)| =$$

$$= |A^{-1}(-I_3)(A - I_3)| = |A^{-1}||-I_3||A - I_3| = |A'|(-1)^3|A - I_3| =$$

$$= -|A - I_3|.$$

Thus $2|A - I_3| = 0$. Hence $|A - I_3| = 0$. Thus $(A - I_3)\boldsymbol{x} = \boldsymbol{0}$ has a non-zero solution. Hence $A\boldsymbol{x} = \boldsymbol{x}$ for some $\boldsymbol{x} \neq \boldsymbol{0}$.

Thus line L given by $\boldsymbol{r} = \lambda \boldsymbol{x}$ is fixed pointwise. Hence A preserves distances and fixes every point of L.

Thus A must be a rotation about L. $A\boldsymbol{x} = \boldsymbol{x}$ has the solution

$$\boldsymbol{x} = \begin{pmatrix} \sqrt{3} \\ 1 \\ 1 \end{pmatrix}. \text{ Thus } L \text{ is } \boldsymbol{r} = \lambda \begin{pmatrix} \sqrt{3} \\ 1 \\ 1 \end{pmatrix}.$$

Take vector $\boldsymbol{v} = \begin{pmatrix} 0 \\ -1 \\ 1 \end{pmatrix}$. This is perpendicular to the axis of rotation L. The rotation sends $\boldsymbol{v}$ into

$$\boldsymbol{v}' = A\boldsymbol{v} = \begin{pmatrix} \frac{\sqrt{3}}{4} \\ \frac{1}{2} \\ \frac{-5}{4} \end{pmatrix}$$

If θ is the angle of rotation, we have: $\quad \boldsymbol{v}' \cdot \boldsymbol{v} = |\boldsymbol{v}'||\boldsymbol{v}| \cos\theta.$

$$\text{Thus } \cos\theta = \frac{\boldsymbol{v}' \cdot \boldsymbol{v}}{|\boldsymbol{v}'||\boldsymbol{v}|} = \frac{-\frac{7}{4}}{\sqrt{2}\sqrt{2}} = -\frac{7}{8}$$

Hence $\quad \theta = \cos^{-1}(-\frac{7}{8}).$

Exercises 6

1. (a) $x^2 + 2y^2 + 3z^2 + 6xy + 8yz$

$$= (x + 3y)^2 - 7y^2 + 3z^2 + 8yz$$

$$= (x + 3y)^2 + 3(z^2 - \tfrac{7}{3}y^2 + \tfrac{8}{3}yz)$$

$$= (x + 3y)^2 + 3[(z + \tfrac{4y}{3})^2 - \tfrac{37y^2}{9}]$$

$$= (x + 3y)^2 + 3(z + \tfrac{4y}{3})^2 - \tfrac{37}{3}y^2 = x'^2 + 3y'^2 - \tfrac{37}{3}z'^2,$$

where $\begin{aligned} x' &= x + 3y \\ y' &= \tfrac{4}{3}y + z \\ z' &= y \end{aligned}$ Then $\begin{pmatrix} x' \\ y' \\ z' \end{pmatrix} = \begin{pmatrix} 1 & 3 & 0 \\ 0 & \frac{4}{3} & 1 \\ 0 & 1 & 0 \end{pmatrix} \begin{pmatrix} x \\ y \\ z \end{pmatrix}.$

The matrix T of the transformation is **not** orthogonal, but is non-singular since $|T| \neq 0$. Rank is 3. Signature $2 - 1 = 1$.

(b) $x^2 + 5y^2 + 3z^2 + 2xy + 4xz = (x + y + 2z)^2 + 4y^2 - z^2 - 4yz =$

$$= (x + y + 2z)^2 + (2y - z)^2 - 2z^2.$$

Put $\begin{aligned} x' &= x + y + 2z \\ y' &= 2y - z \\ z' &= z \end{aligned}$ Then $\begin{pmatrix} x' \\ y' \\ z' \end{pmatrix} = \begin{pmatrix} 1 & 1 & 2 \\ 0 & 2 & -1 \\ 0 & 0 & 1 \end{pmatrix} \begin{pmatrix} x \\ y \\ z \end{pmatrix}.$

$|T| \neq 0$, so T is non-singular. T is not orthogonal. Rank 3, signature 1.

(c) $Q = xy + zy + zx.$ Put $\begin{aligned} x &= x_1 + y_1 \\ y &= x_1 - y_1. \end{aligned}$

Then $Q = (x_1 + y_1)(x_1 - y_1) + z\,(2x_1)$

$$= x_1^2 - y_1^2 + 2x_1 z = (x_1 + z)^2 - y_1^2 - z^2.$$

Put $x' = x_1 + z = (x+y)/2 + z$

$y' = y_1 \quad = (x-y)/2$

$z' = z.$

Thus $$\begin{pmatrix} x' \\ y' \\ z' \end{pmatrix} = \begin{pmatrix} \frac{1}{2} & \frac{1}{2} & 1 \\ \frac{1}{2} & -\frac{1}{2} & 0 \\ 0 & 0 & 1 \end{pmatrix} \begin{pmatrix} x \\ y \\ z \end{pmatrix} = T \begin{pmatrix} x \\ y \\ z \end{pmatrix}.$$

$Q = x'^2 - y'^2 - z'^2$. The matrix T is not orthogonal, but is non-singular because $|T| \neq 0$. Rank 3, signature $1 - 2 = -1$.

2. (a) $x^2 + 5y^2 + 3z^2 + 8yz - 8xz = (x, y, z) \begin{pmatrix} 1 & 0 & -4 \\ 0 & 5 & 4 \\ -4 & 4 & 3 \end{pmatrix} \begin{pmatrix} x \\ y \\ z \end{pmatrix} =$

$= \boldsymbol{x}'A\boldsymbol{x}$. $0 = |A - \lambda I_3| = \begin{vmatrix} 1-\lambda & 0 & -4 \\ 0 & 5-\lambda & 4 \\ -4 & 4 & 3-\lambda \end{vmatrix}$, gives

$$0 = \lambda^3 - 9\lambda^2 - 9\lambda + 81 = (\lambda - 3)(\lambda + 3)(\lambda - 9).$$

Eigenvalues of A are $\lambda_1 = 3, \ \lambda_2 = -3, \ \lambda_3 = 9.$

Let eigenvector belonging to $\lambda_1 = 3$ be $\boldsymbol{x}_1$, then $(A - \lambda_1 I_3)\boldsymbol{x}_1 = \boldsymbol{0}.$

This gives the set of linear homogeneous equations for $\boldsymbol{x}_1 = \begin{pmatrix} a_1 \\ a_2 \\ a_3 \end{pmatrix}$:

$$\begin{aligned} -2a_1 + 0a_2 - 4a_3 &= 0 \\ 0a_1 + 2a_2 + 4a_3 &= 0 \\ -4a_1 + 4a_2 + 0a_3 &= 0 \end{aligned}$$

Solve in the usual way. Sometimes, as here, a solution can be written down by inspection.

Take $a_1 = 2,\ a_2 = 2,\ a_3 = -1$. Now normalise and get

$$\boldsymbol{x}_1 = \tfrac{1}{3}\begin{pmatrix} 2 \\ 2 \\ -1 \end{pmatrix}.$$

Similarly $\boldsymbol{x}_2 = \tfrac{1}{3}\begin{pmatrix} 2 \\ -1 \\ 2 \end{pmatrix}$ belongs to $\lambda_2 = -3$.

$\boldsymbol{x}_3 = \tfrac{1}{3}\begin{pmatrix} 1 \\ -2 \\ -2 \end{pmatrix}$ belongs to $\lambda_3 = 9$.

Take transformation:

$$\begin{pmatrix} x \\ y \\ z \end{pmatrix} = \tfrac{1}{3}\begin{pmatrix} 2 & 2 & 1 \\ 2 & -1 & -2 \\ -1 & 2 & -2 \end{pmatrix}\begin{pmatrix} X \\ Y \\ Z \end{pmatrix} = T'\begin{pmatrix} X \\ Y \\ Z \end{pmatrix}.$$

The matrix T' of this transformation is orthogonal and $|T'| = 1$. 'Solving' the above transformation, using $T^{-1} = T'$, we obtain

$\begin{pmatrix} X \\ Y \\ Z \end{pmatrix}$ in terms of $\begin{pmatrix} x \\ y \\ z \end{pmatrix}$, namely:

$$\begin{pmatrix} X \\ Y \\ Z \end{pmatrix} = T\begin{pmatrix} x \\ y \\ z \end{pmatrix} = \tfrac{1}{3}\begin{pmatrix} 2 & 2 & -1 \\ 2 & -1 & 2 \\ 1 & -2 & -2 \end{pmatrix}\begin{pmatrix} x \\ y \\ z \end{pmatrix}.$$

This expresses the new variables X, Y, Z in terms of the old x, y, z. If we regarded the x, y, z as the usual Cartesian axes, then T would give a 'rotation of axes' to the new coordinates X, Y, Z.

We have: $x^2 + 5y^2 + 3z^2 + 8yz - 8xz = \boldsymbol{x}'A\,\boldsymbol{x} =$

$$= X'TA\,T'X = (X, Y, Z)\ \tfrac{1}{3}\begin{pmatrix} 2 & 2 & -1 \\ 2 & -1 & 2 \\ 1 & -2 & -2 \end{pmatrix}\begin{pmatrix} 1 & 0 & -4 \\ 0 & 5 & 4 \\ -4 & 4 & 3 \end{pmatrix}$$

$$\tfrac{1}{3}\begin{pmatrix} 2 & 2 & 1 \\ 2 & -2 & -2 \\ -1 & 2 & -2 \end{pmatrix}\begin{pmatrix} X \\ Y \\ Z \end{pmatrix}$$

$$= (X, Y, Z)\begin{pmatrix} 3 & 0 & 0 \\ 0 & -3 & 0 \\ 0 & 0 & 9 \end{pmatrix}\begin{pmatrix} X \\ Y \\ Z \end{pmatrix} = 3X^2 - 3Y^2 + 9Z^2.$$

This is the expected form with the eigenvalues as coefficients. Note that if the reduction is not orthogonal, then the coefficients are not necessarily the eigenvalues of the matrix of the original form.

(b) Eigenvalues are $\lambda_1 = 3,\ \lambda_2 = 6,\ \lambda_3 = 9.$

Normalised vectors: $\tfrac{1}{3}\begin{pmatrix} 2 \\ 2 \\ -1 \end{pmatrix},\ \tfrac{1}{3}\begin{pmatrix} 2 \\ -1 \\ 2 \end{pmatrix},\ \tfrac{1}{3}\begin{pmatrix} 1 \\ -2 \\ -2 \end{pmatrix}.$

Transformation is: $\begin{pmatrix} x \\ y \\ z \end{pmatrix} = \tfrac{1}{3}\begin{pmatrix} 2 & 2 & 1 \\ 2 & -1 & -2 \\ -1 & 2 & -2 \end{pmatrix}\begin{pmatrix} X \\ Y \\ Z \end{pmatrix} = T'\begin{pmatrix} X \\ Y \\ Z \end{pmatrix},$

where, as before, T' is the transpose of T.

$$T' = \tfrac{1}{3}\begin{pmatrix} 2 & 2 & 1 \\ 2 & -1 & -2 \\ -1 & 2 & -2 \end{pmatrix}$$ is orthogonal and $|T'| = 1$.

$$\begin{pmatrix} X \\ Y \\ Z \end{pmatrix} = T\begin{pmatrix} x \\ y \\ z \end{pmatrix} = \tfrac{1}{3}\begin{pmatrix} 2 & 2 & -1 \\ 2 & -1 & 2 \\ 1 & -2 & -2 \end{pmatrix}\begin{pmatrix} x \\ y \\ z \end{pmatrix}.$$

Quadratic form becomes:

$$(X, Y, Z)\, T A T' \begin{pmatrix} X \\ Y \\ Z \end{pmatrix} = (X, Y, Z) \begin{pmatrix} 3 & 0 & 0 \\ 0 & 6 & 0 \\ 0 & 0 & 9 \end{pmatrix} \begin{pmatrix} X \\ Y \\ Z \end{pmatrix}.$$

$$= 3X^2 + 6Y^2 + 9Z^2.$$

(c) $5x^2 + 11y^2 - 2z^2 + 12xz + 12yz = \mathbf{x}' \begin{pmatrix} 5 & 0 & 6 \\ 0 & 11 & 6 \\ 6 & 6 & -2 \end{pmatrix} \mathbf{x} = \mathbf{x}' A\, \mathbf{x}.$

Eigenvalues: $\lambda_1 = 7,\ \lambda_2 = -7,\ \lambda_3 = 14.$

Normalised eigenvectors: $\frac{1}{7}\begin{pmatrix} 6 \\ -3 \\ 2 \end{pmatrix},\ \frac{1}{7}\begin{pmatrix} 3 \\ 2 \\ -6 \end{pmatrix},\ \frac{1}{7}\begin{pmatrix} 2 \\ 6 \\ 3 \end{pmatrix}.$

$$T' = \frac{1}{7}\begin{pmatrix} 6 & 3 & 2 \\ -3 & 2 & 6 \\ 2 & -6 & 3 \end{pmatrix}. \qquad T = \frac{1}{7}\begin{pmatrix} 6 & -3 & 2 \\ 3 & 2 & -6 \\ 2 & 6 & 3 \end{pmatrix}.$$

T' and T are orthogonal and $|T| = 1$.

$$\begin{pmatrix} x \\ y \\ z \end{pmatrix} = T' \begin{pmatrix} X \\ Y \\ Z \end{pmatrix}, \quad \begin{pmatrix} X \\ Y \\ Z \end{pmatrix} = T \begin{pmatrix} x \\ y \\ z \end{pmatrix} \quad \text{lead to}$$

$$\mathbf{x}' A\, \mathbf{x} = X' T A T' X = 7X^2 - 7Y^2 + 14Z^2.$$

3. Associated matrix is: $A = \begin{pmatrix} 3 & 1 & 1 \\ 1 & 3 & 1 \\ 1 & 1 & 5 \end{pmatrix}.$

$|A - \lambda I_3| = 0$ gives eigenvalues $\lambda_1 = 2,\ \lambda_2 = 3,\ \lambda_3 = 6,$

and normalised eigenvectors: $\frac{1}{\sqrt{2}}\begin{pmatrix} 1 \\ -1 \\ 0 \end{pmatrix},\ \frac{1}{\sqrt{3}}\begin{pmatrix} 1 \\ 1 \\ -1 \end{pmatrix},\ \frac{1}{\sqrt{6}}\begin{pmatrix} 1 \\ 1 \\ 2 \end{pmatrix}.$

Take $T' = \begin{pmatrix} \frac{1}{\sqrt{2}} & \frac{1}{\sqrt{3}} & \frac{1}{\sqrt{6}} \\ \frac{-1}{\sqrt{2}} & \frac{1}{\sqrt{3}} & \frac{1}{\sqrt{6}} \\ 0 & \frac{-1}{\sqrt{3}} & \frac{2}{\sqrt{6}} \end{pmatrix}$. Put $\begin{pmatrix} x_1 \\ x_2 \\ x_3 \end{pmatrix} = T' \begin{pmatrix} y_1 \\ y_2 \\ y_3 \end{pmatrix}$.

Then $y = Tx$ is the required rotation of axes, since T is orthogonal and $|T| = 1$.

Quadric in the new axes is $2y_1^2 + 3y_2^2 + 6y_3^2 = 1$.

This is: $\dfrac{y_1^2}{\left(\frac{1}{\sqrt{2}}\right)^2} + \dfrac{y_2^2}{\left(\frac{1}{\sqrt{3}}\right)^2} + \dfrac{y_3^2}{\left(\frac{1}{\sqrt{6}}\right)^2} = 1$ which is an ellipsoid with

semi-axes of lengths $\frac{1}{\sqrt{2}}, \quad \frac{1}{\sqrt{3}}, \quad \frac{1}{\sqrt{6}},$

and with principal axes y_1, y_2, y_3 in directions given by the eigenvectors. Thus y_1 axis has direction cosines

$\frac{1}{\sqrt{2}}, \quad \frac{-1}{\sqrt{2}}, \quad 0;$ y_2 axis $\frac{1}{\sqrt{3}}, \quad \frac{1}{\sqrt{3}}, \quad \frac{-1}{\sqrt{3}};$

and y_3 axis $\frac{1}{\sqrt{6}}, \quad \frac{1}{\sqrt{6}}, \quad \frac{2}{\sqrt{6}}$.

4. Matrix of the quadratic form is: $A = \begin{pmatrix} 2 & 4 & -2 \\ 4 & 2 & -2 \\ -2 & -2 & -1 \end{pmatrix}$

Solving $|A - \lambda I| = 0$, which is $\begin{vmatrix} 2-\lambda & 4 & -2 \\ 4 & 2-\lambda & -2 \\ -2 & -2 & -1-\lambda \end{vmatrix} = 0,$

gives eigenvalues: $\lambda_1 = -2, \ \lambda_2 = -2, \ \lambda_3 = 7.$

Corresponding normalised eigenvectors are:

$$\frac{1}{\sqrt{2}}\begin{pmatrix}-1\\1\\0\end{pmatrix}, \quad \frac{\sqrt{2}}{6}\begin{pmatrix}1\\1\\4\end{pmatrix}, \quad \frac{1}{3}\begin{pmatrix}2\\2\\-1\end{pmatrix}.$$

Note that the first two eigenvectors both belong to the eigenvalue -2. This is an example of a repeated root. We get two linearly independent eigenvectors belonging to -2, whose eigenspace is of dimension 2. By taking suitable linear combinations of these we get two orthogonal eigenvectors for $\lambda_1 = \lambda_2 = -2$. We change coordinates to $\boldsymbol{x} = T'\boldsymbol{y}$,

$$\text{where } T' = \begin{pmatrix} \frac{-1}{\sqrt{2}} & \frac{\sqrt{2}}{6} & \frac{2}{3} \\ \frac{1}{\sqrt{2}} & \frac{\sqrt{2}}{6} & \frac{2}{3} \\ 0 & \frac{2\sqrt{2}}{3} & \frac{-1}{3} \end{pmatrix}, \text{ and } |T'| = 1.$$

In the new coordinates the surface has the equation:

$$-2y_1^2 - 2y_2^2 + 7y_3^2 = 1.$$

This is an hyperboloid of two sheets with semi-axes of lengths

$$\frac{1}{\sqrt{2}}, \quad \frac{1}{\sqrt{2}}, \quad \frac{1}{\sqrt{7}},$$

as can be seen by writing it in the form:

$$-\frac{y_1^2}{\left(\frac{1}{\sqrt{2}}\right)^2} - \frac{y_2^2}{\left(\frac{1}{\sqrt{2}}\right)^2} + \frac{y_3^2}{\left(\frac{1}{\sqrt{7}}\right)^2} = 1.$$

The axis of symmetry is the y_3 axis. Hence we have a quadric of revolution. In fact this was shown immediately by the repeated eigenvalue -2.

The direction of the y_3 axis is given by the third column of T', that is by the eigenvector belonging to $\lambda_3 = 7$. Thus the direction cosines of the axis of symmetry are $\frac{2}{3}, \frac{2}{3}, \frac{-1}{3}$ relative to the original $0x_1\,x_2\,x_3$ coordinate system.

5. Associated matrix is $A = \begin{pmatrix} 7 & -4 & 0 \\ -4 & 5 & 4 \\ 0 & 4 & 3 \end{pmatrix}$.

Eigenvalues are: $\lambda_1 = -1,\ \lambda_2 = 5,\ \lambda_3 = 11.$

Normalised eigenvectors are: $\frac{1}{3}\begin{pmatrix} 1 \\ 2 \\ -2 \end{pmatrix},\ \frac{1}{3}\begin{pmatrix} 2 \\ 1 \\ 2 \end{pmatrix},\ \frac{1}{3}\begin{pmatrix} 2 \\ -2 \\ -1 \end{pmatrix}.$

Coordinate transformation is:

$$x = T'y, \text{ where } T' = \frac{1}{3}\begin{pmatrix} 1 & 2 & 2 \\ 2 & 1 & -2 \\ -2 & 2 & -1 \end{pmatrix},\ |T'| = 1.$$

Equation of surface: $-y_1^2 + 5y_2^2 + 11y_3^2 = 1.$

This is an hyperboloid of one sheet. Relative to the original $0x_1\,x_2\,x_3$ axes, the y_1 axis has direction cosines

$\frac{1}{3}, \frac{2}{3}, \frac{-2}{3}$, y_2 axis $\frac{2}{3}, \frac{1}{3}, \frac{2}{3}$, and y_3 axis $\frac{2}{3}, \frac{-2}{3}, \frac{-1}{3}$.

These directions are given by the eigenvectors.

6. Let centre be (a,b,c). Change coordinates so centre is new origin.

$$\begin{aligned} x &= X + a \\ y &= Y + b \\ z &= Z + c. \end{aligned}$$

Equate coefficients of X, Y, Z to zero to get equations for a,b,c:

$$\begin{aligned} 8a + 8 &= 0 \\ 18b + 4c + 40 &= 0 \\ 12c + 4b + 20 &= 0 \end{aligned}$$

Thus centre is: $(-1, -2, -1)$.

The change of coordinates does not affect the matrix of the quadratic form part of the equation of the quadric.

In fact the equation becomes:

$$4X^2 + 9Y^2 + 6Z^2 + 4YZ = 20.$$

This matrix is: $A = \begin{pmatrix} 4 & 0 & 0 \\ 0 & 9 & 2 \\ 0 & 2 & 6 \end{pmatrix}$.

$$0 = |A - \lambda I_3| = (\lambda - 4)(\lambda - 5)(\lambda - 10).$$

Thus eigenvalues are 4, 5, 10. The quadric is an ellipsoid.

Normalised eigenvectors are: $\begin{pmatrix} 1 \\ 0 \\ 0 \end{pmatrix}, \begin{pmatrix} 0 \\ \frac{1}{\sqrt{5}} \\ \frac{-2}{\sqrt{5}} \end{pmatrix}, \begin{pmatrix} 0 \\ \frac{2}{\sqrt{5}} \\ \frac{1}{\sqrt{5}} \end{pmatrix}$.

Hence direction cosines of the principal axes are:

$$1, 0, 0;\ 0, \frac{1}{\sqrt{5}}, \frac{-2}{\sqrt{5}};\ 0, \frac{2}{\sqrt{5}}, \frac{1}{\sqrt{5}}$$

relative to the original axes. They are given by the eigenvectors. Changing from the axes X, Y, Z to axes x_1, x_2, x_3 by

$$\begin{pmatrix} X \\ Y \\ Z \end{pmatrix} = T' \begin{pmatrix} x_1 \\ x_2 \\ x_3 \end{pmatrix}$$, where T', the transpose of T, is

$$T' = \begin{pmatrix} 1 & 0 & 0 \\ 0 & \frac{1}{\sqrt{5}} & \frac{2}{\sqrt{5}} \\ 0 & \frac{-2}{\sqrt{5}} & \frac{1}{\sqrt{5}} \end{pmatrix}$$, we get the equation of the quadric:

$$\frac{x_1^2}{(\sqrt{5})^2} + \frac{x_2^2}{(\sqrt{4})^2} + \frac{x_3^2}{(\sqrt{2})^2} = 1.$$

Thus lengths of the principal axes are:

$$2\sqrt{5},\ 4,\ 2\sqrt{2}.$$

7. Associated matrices of the two forms are:

$$A = \begin{pmatrix} 1 & -3 & -1 \\ -3 & 5 & 3 \\ -1 & 3 & 3 \end{pmatrix}, \qquad B = \begin{pmatrix} 2 & 0 & 1 \\ 0 & 6 & 5 \\ 1 & 5 & 5 \end{pmatrix}.$$

Consider $|A - \lambda B| = 0$. This gives:

$$0 = \begin{vmatrix} 1-2\lambda & -3 & -1-\lambda \\ -3 & 5-6\lambda & 3-5\lambda \\ -1-\lambda & 3-5\lambda & 3-5\lambda \end{vmatrix} = \begin{vmatrix} 1-2\lambda & -3 & -1-\lambda \\ -3 & 5-6\lambda & 3-5\lambda \\ 2-\lambda & \lambda-2 & 0 \end{vmatrix}.$$

$$0 = (\lambda - 2) \begin{vmatrix} 1-2\lambda & -3 & -1-\lambda \\ -3 & 5-6\lambda & 3-5\lambda \\ -1 & 1 & 0 \end{vmatrix}.$$ This reduces to:

$(\lambda - 2)(\lambda - 1)(\lambda + 1) = 0.$ Roots are $\lambda_1 = 1,\ \lambda_2 = -1,\ \lambda_3 = 2.$

For each root λ_i solve $(A - \lambda_i B)\begin{pmatrix} x_1 \\ x_2 \\ x_3 \end{pmatrix} = \begin{pmatrix} 0 \\ 0 \\ 0 \end{pmatrix}$ for $\begin{pmatrix} x_1 \\ x_2 \\ x_3 \end{pmatrix}$; $i = 1,2,3$.

No need to normalise, as we are not looking for orthogonal matrix. Suitable right factors of zero are:

$$\begin{pmatrix} 1 \\ 1 \\ -2 \end{pmatrix}, \begin{pmatrix} 1 \\ 1 \\ -1 \end{pmatrix}, \begin{pmatrix} 0 \\ 1 \\ -1 \end{pmatrix}. \qquad \text{Put } T = \begin{pmatrix} 1 & 1 & 0 \\ 1 & 1 & 1 \\ -2 & -1 & -1 \end{pmatrix}.$$

Change variables according to:

$$\begin{pmatrix} x \\ y \\ z \end{pmatrix} = T \begin{pmatrix} X \\ Y \\ Z \end{pmatrix}. \qquad \begin{aligned} x &= X + Y \\ y &= X + Y + Z \\ z &= -2X - Y - Z. \end{aligned}$$

Since $|T| \neq 0$, we can find T^{-1} and solve for X, Y, Z if we wish. The two forms reduce to:

$$(X, Y, Z)\ T'A\ T \begin{pmatrix} X \\ Y \\ Z \end{pmatrix} = 4X^2 - Y^2 + 2Z^2 \text{ and}$$

$$(X, Y, Z)\ T'B\ T \begin{pmatrix} X \\ Y \\ Z \end{pmatrix} = 4X^2 + Y^2 + Z^2 \text{ respectively.}$$

8. (a) Matrix of form is:

$$A = \begin{pmatrix} 3 & -3 & -2 \\ -3 & 5 & 4 \\ -2 & 4 & 7 \end{pmatrix}$$

We have the sequence of determinants:

$$3, \begin{vmatrix} 3 & -3 \\ -3 & 5 \end{vmatrix} = 6, \quad |A| = 22.$$

These are all positive, so form is positive definite. The rank is 3 and signature is also 3, since a positive definite form must reduce to:

$$\mu_1 X^2 + \mu_2 Y^2 + \mu_3 Z^2, \text{ with } \mu_1 > 0,\ \mu_2 > 0,\ \mu_3 > 0.$$

(b) Associated matrix is:

$$A = \begin{pmatrix} 1 & 1 & 1 \\ 1 & 1 & 1 \\ 1 & 1 & 1 \end{pmatrix}.$$

Consider sequence of determinants:

$$1, \begin{vmatrix} 1 & 1 \\ 1 & 1 \end{vmatrix} = 0, \quad |A| = 0.$$

Thus form is **not** positive definite. In fact the form can be written:

$$(x + y + z)^2$$

If we put $\begin{matrix} X & = & x + y + z \\ Y & = & y \\ Z & = & z \end{matrix}$ we have:

$$\begin{pmatrix} X \\ Y \\ Z \end{pmatrix} = \begin{pmatrix} 1 & 1 & 1 \\ 0 & 1 & 0 \\ 0 & 0 & 1 \end{pmatrix} \begin{pmatrix} x \\ y \\ z \end{pmatrix}$$

and the transformation is non-singular. The form reduces to just X^2.

Rank is 1, and signature is 1. The rank is also the rank of A, which can be seen at a glance to be 1.

(c) Matrix of form is:

$$A = \begin{pmatrix} 1 & -1 & -1 \\ -1 & 1 & -1 \\ -1 & -1 & 1 \end{pmatrix}$$

Sequence of determinants is:

$$1, \begin{vmatrix} 1 & -1 \\ -1 & 1 \end{vmatrix} = 0, \quad |A| = -4.$$

Form **not** positive definite. Form can be reduced to

$$X^2 - 4Y^2 + 4Z^2, \text{ where } \begin{cases} X = x - y - z \\ Y = \dfrac{y}{2} + \dfrac{z}{2} \\ Z = \dfrac{y}{2} - \dfrac{z}{2} \end{cases}.$$

Rank is 3 and signature is 1.

(d) Matrix of form is:

$$A = \begin{pmatrix} 2 & 0 & 1 \\ 0 & 2 & 1 \\ 1 & 1 & 1 \end{pmatrix}$$

Sequence of determinants is:

$$2, \quad \begin{vmatrix} 2 & 0 \\ 0 & 2 \end{vmatrix} = 4, \quad |A| = 0.$$

Form is **not** positive definite. In fact this is immediate from rank of form = rank of $A = 2$.

The form can be reduced to:

$$(x + y + z)^2 + (x - y)^2 =$$

$$= X^2 + Y^2, \text{ where } \quad X = x + y + z, \quad Y = x - y, \quad Z = z.$$

Thus rank is 2 and signature is 2.

9. Write kinetic energy $T = \frac{m}{2} F$ and $V = mg\, G$.

The matrix associated with F is: $A = \begin{pmatrix} 3 & -1 \\ -1 & 3 \end{pmatrix}$ and

the matrix $B = \begin{pmatrix} 5 & 1 \\ 1 & 5 \end{pmatrix}$ is associated with G.

Consider $|A - \lambda B| = 0.$ This is: $\begin{vmatrix} 3-5\lambda & -1-\lambda \\ -1-\lambda & 3-5\lambda \end{vmatrix} = 0.$

This reduces to: $(3\lambda - 1)(\lambda - 1) = 0.$ Roots are: $\lambda_1 = \frac{1}{3}, \ \lambda_2 = 1.$

$$(A - \tfrac{1}{3}B) \begin{pmatrix} x_1 \\ x_2 \end{pmatrix} = \begin{pmatrix} 0 \\ 0 \end{pmatrix} \text{ has solution: } \begin{pmatrix} x_1 \\ x_2 \end{pmatrix} = \begin{pmatrix} 1 \\ 1 \end{pmatrix}.$$

$$(A - B) \begin{pmatrix} x_1 \\ x_2 \end{pmatrix} = \begin{pmatrix} 0 \\ 0 \end{pmatrix} \text{ has solution: } \begin{pmatrix} x_1 \\ x_2 \end{pmatrix} = \begin{pmatrix} 1 \\ -1 \end{pmatrix}.$$

Put $P = \begin{pmatrix} 1 & 1 \\ 1 & -1 \end{pmatrix}$. Change variables according to: $\begin{pmatrix} q_1 \\ q_2 \end{pmatrix} = P \begin{pmatrix} y_1 \\ y_2 \end{pmatrix}.$

$$F \text{ becomes } (\dot{y}_1, \dot{y}_2) \quad P' \begin{pmatrix} 3 & -1 \\ -1 & 3 \end{pmatrix} \quad P \begin{pmatrix} \dot{y}_1 \\ \dot{y}_2 \end{pmatrix} = 4\dot{y}_1^2 + 8\dot{y}_2^2.$$

$$G \text{ becomes } (y_1, y_2) \quad P' \begin{pmatrix} 5 & 1 \\ 1 & 5 \end{pmatrix} \quad P \begin{pmatrix} y_1 \\ y_2 \end{pmatrix} = 12y_1^2 + 8y_2^2.$$

Thus $T = \frac{m}{2}(4\dot{y}_1^2 + 8\dot{y}_2^2)$, $V = mg\,(12y_1^2 + 8y_2^2)$, in normal coordinates y_1 and y_2.

From $\begin{pmatrix} q_1 \\ q_2 \end{pmatrix} = P \begin{pmatrix} y_1 \\ y_2 \end{pmatrix}$ we get:
$$\begin{aligned} q_1 &= y_1 + y_2 \\ q_2 &= y_1 - y_2\,. \end{aligned}$$

Solving we get:
$$\begin{aligned} y_1 &= (q_1 + q_2)/2 \\ y_2 &= (q_1 - q_2)/2 \end{aligned}$$

The Lagrange equations for y_1 and y_2 lead to the differential equations:

$$\ddot{y}_1 + 6g\,y_1 = 0$$

$$\ddot{y}_2 + 2g\,y_2 = 0$$

These are simple harmonic motion type equations with general solution:

$$y_1 = a_1 \sin(\sqrt{6g}\,t + b_1)$$

$$y_2 = a_2 \sin(\sqrt{2g}\,t + b_2),$$

where a_1, a_2, b_1, b_2 are arbitrary constants.

The periods of the normal modes are:

$$\frac{2\pi}{\sqrt{6g}}, \qquad \frac{2\pi}{\sqrt{2g}}$$

One normal mode occurs when $a_1 \neq 0$, $a_2 = 0$

and the second when $a_1 = 0$, $a_2 \neq 0$.

Exercises 7

1. $$\begin{aligned} 1079 &= 741 + 338, \\ 741 &= 2(338) + 65, \\ 338 &= 5(65) + 13, \\ 65 &= 5(13). \end{aligned}$$

Thus G.C.D. of 741 and 1079 is 13.

Moreover: $$\begin{aligned} 13 &= 338 - 5(65) = 338 - 5(741 - 2(338)) \\ &= 11(338) - 5(741) = 11(1079 - 741) - 5(741). \end{aligned}$$

Thus $$13 = 11(1079) - 16(741).$$

2. (a) $$\begin{aligned} x^4 - x^3 + x - 1 &= x(x^3 - x^2 + x - 1) + (-x^2 + 2x - 1) \\ x^3 - x^2 + x - 1 &= (-x - 1)(-x^2 + 2x - 1) + 2(x - 1) \\ -x^2 + 2x - 1 &= (\tfrac{-x}{2} + \tfrac{1}{2})\, 2(x - 1). \end{aligned}$$

Thus a G.C.D. of $x^4 - x^3 + x - 1$ and $x^3 - x^2 + x - 1$ is $2(x - 1)$.

Note that $k(x - 1)$ is also a G.C.D. in $\mathbf{Q}[x]$, where $k \in \mathbf{Q}$.

We have:

$$\begin{aligned} 2(x - 1) &= (x^3 - x^2 + x - 1) + (x + 1)(-x^2 + 2x - 1) = \\ &= (x^3 - x^2 + x - 1) + (x + 1)\, [(x^4 - x^3 + x - 1) - x(x^3 - x^2 + x - 1 \\ &= (x + 1)(x^4 - x^3 + x - 1) + (-x^2 - x + 1)(x^3 - x^2 + x - 1). \end{aligned}$$

Thus:

$$2(x - 1) = (x + 1)(x^4 - x^3 + x - 1) + (-x^2 - x + 1)(x^3 - x^2 + x - 1).$$

(b) $$\begin{aligned} x^3 + x + 1 &= (x + 1)(x^2 - x + 1) + x \\ x^2 - x + 1 &= (x - 1)x + 1 \\ x &= x(1). \end{aligned}$$

Thus G.C.D. of $x^3 + x + 1$ and $x^2 - x + 1$ is 1 or in fact any $k \in \mathbf{Q}$. We have:

$$\begin{aligned} 1 &= (x^2 - x + 1) - (x - 1)x = \\ &= (x^2 - x + 1) - (x - 1)[(x^3 + x + 1) - (x + 1)(x^2 - x + 1)] \end{aligned}$$

$$= -(x-1)(x^3+x+1) + x^2(x^2-x+1).$$

Thus $1 = x^2(x^2-x+1) + (1-x)(x^3+x+1).$

(c) $x^4+x^2+x+1 = 1\cdot(x^4+1) + (x^2+x)$

$x^4+1 = (x^2-x+1)(x^2+x) + (-x+1).$

$x^2+x = (-x-2)(-x+1) + 2$

$(-x+1) = (\frac{-x}{2} + \frac{1}{2})\,2.$

Thus G.C.D. is 2, or any $k \in \mathbf{Q}$.

We have:

$$2 = (-x^3-x^2+x-1)(x^4+x^2+x+1) + (x^3+x^2+3)(x^4+1).$$

(d) $x^4-x^3+3x^2-2x+2 = 1\cdot(x^4+x^2-2) + (-x^3+2x^2-2x+4)$

$x^4+x^2-2 = (-x-2)(-x^3+2x^2-2x+4) + 3(x^2+2)$

$-x^3+2x^2-2x+4 = \frac{1}{3}(-x+2)(3x^2+6).$

Thus G.C.D. of $x^4-x^3+3x^2-2x+2$ and x^4+x^2-2 is $k(x^2+2)$, $k \in \mathbf{Q}$.

$$3(x^2+2) = (x+2)(x^4-x^3+3x^2-2x+2) + (-x-1)(x^4+x^2-2).$$

(e) $x^5+x+1 = x(x^4+x^2+2) + (-x^3-x+1)$

$x^4+x^2+2 = -x(-x^3-x+1) + (x+2)$

$-x^3-x+1 = (-x^2+2x-5)(x+2) + 11.$

$x+2 = \frac{1}{11}(x+2)\cdot 11.$

Thus any $k \in \mathbf{Q}$ is a G.C.D. of x^5+x+1 and x^4+x^2+2. We have:

$$11 = (x^3-2x^2+5x+1)(x^5+x+1) + (-x^4+2x^3-4x^2-3x+5)(x^4+x^2+2).$$

3. $x = 1 + 5i, \quad y = 4 + 2i,$

$$\frac{x}{y} = \frac{7}{10} + \left(\frac{9}{10}\right) i.$$

Nearest Gaussian integer to $\frac{x}{y}$ on Argand diagram is:

$1 + i.$ Take $q = 1 + i.$

Then $x = (1 + i)\, y + r.$

Thus $(1 + 5i) = (1 + i)(4 + 2i) + (-1 - i)$, where $r = (-1 - i)$.

$\delta(r) = 2, \quad \delta(y) = 20.$ Thus $\delta(r) < \delta(y)$.

Repeat for: $(4 + 2i) = q_1(-1 - i) + r_1.$

We have: $(4 + 2i)/(-1 - i) = -3 + i$, which is a Gaussian integer.

Thus $(4 + 2i) = (-3 + i)(-1 - i) + 0.$

Hence a G.C.D of $1 + 5i$ and $4 + 2i$ is $(-1 - i)$ or any Gaussian integer of the form $u(-1 - i)$, where u is a unit in $\mathbf{Z}[i]$. We have:

$(-1 - i) = (1 + 5i) + (-1 - i)(4 + 2i)$, which is in the form

$A(1 + 5i) + B(4 + 2i)$, with $A = 1$ and $B = -(1 + i)$.

If we take $u = -1$, obviously a unit in $\mathbf{Z}[i]$, we could take $1 + i$ as a G.C.D of $1 + 5i$ and $4 + 2i$.

Then $(1 + i) = (-1)(1 + 5i) + (1 + i)(4 + 2i).$

4. (a) Let $x \mid y$ in $\mathbf{Z}[i]$.

Then $y = tx$, for some $t \in \mathbf{Z}[i]$, $t \neq 0$.

Then $\delta(y) = |tx|^2 = |t|^2 |x|^2 = \delta(t)\,\delta(x).$

Thus $\delta(x) \mid \delta(y)$, in $\{0, 1, 2, 3, \ldots\}$.

(b) Let x be unit in $\mathbf{Z}[i]$.

Then $xu = 1$, for some $u \in \mathbf{Z}[i]$.

By (a) above, $x \mid 1$, hence $\delta(x) \mid \delta(1)$.

Now $\quad \delta(1) = \delta(1 + 0i) = 1.$

Thus $\quad \delta(x) \mid 1$ in $\{0, 1, 2, 3, \ldots\}$. Hence $\delta(x) = 1.$

Conversely, suppose $\delta(x) = 1.$

Put $x = a + ib.$ Then $1 = \delta(x) = a^2 + b^2;\ a,b \in \mathbf{Z}.$

Thus $x = 1 + 0i,\ -1 + 0i,\ 0 + i,\ 0 - i.$

Now $1 \cdot 1 = 1,\ (-1)(-1) = 1,\ i(-i) = 1, (-i)\,i = 1.$

Thus all these values of x are units. Hence x is a unit if and only if $\delta(x) = 1.$

It follows that the units of $\mathbf{Z}[i]$ are precisely the 4 complex numbers $1, -1, i, -i.$

Under the multiplication in $\mathbf{Z}[i]$, these form a group of order 4. In fact this group of units is cyclic generated by i. It is:

$$\{i,\ \ i^2 = -1,\ \ i^3 = -i,\ \ i^4 = 1\} = \langle i \rangle.$$

5. Let R be a field. Let $I \neq \{0\}$ be an ideal of R. Then $0 \neq a \in I$. But then $a \cdot a^{-1} \in I$.

Now $a\,a^{-1} = e$, the multiplicative identity of the field. Let b be any element of R. Then, since $e \in I$ and I is an ideal, we have: $b = e\,b \in I.$

Thus $R \subset I$. Obviously $I \subset R$. Hence $I = R$.

Conversely, let R have only the ideals $\{0\}$ and R. Let $a \neq 0$. Let $I = \langle a \rangle$. Then $I \neq \{0\}$ hence $I = R$. Now $e \in R$. Thus $e \in I = \{ra : r \in R\}$. Hence, for some $b \in R$, we have $e = b\,a$. Thus a has a multiplicative inverse. It follows that R is a field.

6. (a) Let $x = a + bi\sqrt{5}, \quad y = c + di\sqrt{5}.$

Then $\quad xy = (ac - 5bd) + (ad + bc)\,i\sqrt{5}.$

Thus
$$\delta(xy) = (ac - 5bd)^2 + 5(ad + bc)^2$$
$$= a^2c^2 - 10\,abcd + 25\,b^2d^2 + 5\,a^2d^2 + 10\,abcd + 5\,b^2c^2.$$

$$\delta(x)\,\delta(y) = (a^2 + 5b^2)(c^2 + 5d^2)$$
$$= a^2c^2 + 5\,b^2c^2 + 5\,a^2d^2 + 25\,b^2d^2.$$

Thus $\quad \delta(xy) = \delta(x)\,\delta(y).$

(b) Let x be unit. Then $xy = 1$, for some $y \in \mathbf{Z}[i\sqrt{5}]$.

Hence $\delta(1) = \delta(xy) = \delta(x)\,\delta(y)$ by (a).

Thus $1 = \delta(x)\,\delta(y)$. Hence $\delta(x) = 1$.

Conversely, let $\delta(x) = 1$. Put $x = a + bi\sqrt{5}$.

Then $a^2 + 5b^2 = 1$. Thus $a = \pm 1,\ b = 0$.

Hence $x = \pm 1$, which is a unit.

Note that this shows that the units of $\mathbf{Z}[i\sqrt{5}]$ are ± 1.

(c) By (b) $\delta(y) > 1$. Thus $\delta(xy) = \delta(x)\,\delta(y) > \delta(x)$.

(d) $\delta(2) = 4$, $\delta(3) = 9$, $\delta(1 + i\sqrt{5}) = \delta(1 - i\sqrt{5}) = 6$.

If $2 = pq$, p and q not units, then $\delta(p) \mid \delta(2)$. Put $p = u + vi\sqrt{5}$.

Then $\delta(p) = u^2 + 5v^2$. Thus $u^2 + 5v^2 = 1, 2$, or 4.

Thus $u = \pm 1, \pm 2$, $v = 0$. Hence $p = \pm 1, \pm 2$.

If $p = \pm 1$, then p is a unit. If $p = \pm 2$, then, since $4 = \delta(2) = \delta(p)\,\delta(q)$,

$\delta(q) = 1$, and q is a unit. But we supposed p and q not units. Hence $2 \neq pq$, with p,q not units. Thus 2 is prime.

Similar arguments show 3, $1 + i\sqrt{5}$, $1 - i\sqrt{5}$ are primes.

(e) If $\mathbf{Z}[i\sqrt{5}]$ is Euclidean, then theorems 7 · 2 · 4 and 7 · 2 · 7 hold.

However: $6 = 2 \times 3 = (1 + i\sqrt{5}) \times (1 - i\sqrt{5})$

contradicts theorem 7 · 2 · 7. Thus $\mathbf{Z}[i\sqrt{5}]$ is not Euclidean.

Further Reading and References

1. Birkhoff, G. and MacLane, S. *'Survey of Modern Algebra'*, Macmillan 4th Edition 1977.
2. Gardiner, C.F. *'A First Course in Group Theory'*, Springer-Verlag 1980.
3. Goult, R.J. *'Applied Linear Algebra'*, Ellis Horwood 1978.
4. Halmos, P.R. *'Naive Set Theory'*, Springer-Verlag 1974.
5. Hartley, B. and Hawkes, T.O. *'Rings, Modules and Linear Algebra'*, Chapman and Hall 1970.
6. Herstein, I.N. *'Topics in Algebra'*, Wiley 1975.
7. Kaplansky, I. *'Linear Algebra and Geometry – a Second Course'*, Chelsea Publishing Company 2nd Edition 1974.
8. Stewart, I. *'Concepts of Modern Mathematics'*, Penguin Books 1975.
9. Stewart, I. and Tall, D.O. *'The Foundations of Mathematics'*, Oxford University Press 1977.

Index

U

V

W

Z